Postural Variations in Childhood

POSTGRADUATE PAEDIATRICS SERIES

under the General Editorship of

JOHN APLEY
C.B.E., M.D., B.S., F.R.C.P., J.P.

Emeritus Consultant Paediatrician, United Bristol Hospitals

Postural Variations in Childhood

CÉCILE ASHER
M.D., M.R.C.P.,
D.C.H., B.Sc.

BUTTERWORTHS
LONDON and BOSTON

THE BUTTERWORTH GROUP

ENGLAND
Butterworth & Co (Publishers) Ltd
London: 88 Kingsway, WC2B 6AB

AUSTRALIA
Butterworth Pty Ltd
Sydney: 586 Pacific Highway, NSW 2067
Melbourne: 343 Little Collins Street. 3000
Brisbane: 240 Queen Street, 4000

CANADA
Butterworth & Co (Canada) Ltd
Toronto: 2265 Midland Avenue,
Scarborough, Ontario, MIP 4Sl

NEW ZEALAND
Butterworths of New Zealand Ltd
Wellington: 26-28 Waring Taylor Street, 1

SOUTH AFRICA
Butterworth & Co (South Africa) (Pty) Ltd
Durban: 152-154 Gale Street

USA
Butterworth
161 Ash Street
Reading, Mass. 01867

First published 1975

ISBN 0 407 00032 1
Suggested UDC No. 612.766: 612.66
Suggested additional no. 616.8–009.18–053.2

Library of Congress Cataloging in Publication Data

Asher, Cecile.
Postural variations in childhood.
(Postgraduate paediatrics series)
Includes bibliographies and index.
1. Children–Growth. 2. Developmental biology. 3. Poster. 1. Title II series.
[DNLM: 1. Postyre–In infancy and childhood. WE103 A825p]
RJ131.A77 612'.7 75–6756
ISBN 0–407–00032–1

Printed in England by J. W. Arrowsmith Ltd of Bristol.

Contents

Preface

This book is concerned with changes in posture which take place during infancy, childhood and adolescence. It is based on a longitudinal (long-term) study of 200 healthy children and should be of interest to those concerned with the supervision and physical welfare of healthy children; to those working in the preventive fields of medicine; to physiotherapists, remedial gymnasts, and specialists in physical education, to health visitors, medical students, and paediatricians.

Postural variations are associated with stages of development; they arise in response to problems of balance which occur as the result of changes in body proportions and body components. Certain postural variations may be regarded as normal. If they occur in certain age groups, they need no treatment, and should not be labelled 'postural defects' as this suggests presence of abnormality.

In this book, a general review of postural development from infancy to adolescence is given, a short account of general growth and development is included, and the natural history of the following conditions is described–knock knee, valgus heel, 'flat foot', metatarsus varus, hallux valgus, tight hamstrings, round shoulders, kyphosis, lordosis, scoliosis. In each case, the anatomy of the region is discussed.

Throughout this study, emphasis is placed on the part played by evolutionary processes in the production of the human child as we know him today. For example, the middle-school child shows traces of his ancestor's tree life in his mobile upper segment, and his versatile use of his arms and shoulder girdle in climbing; adaptation to terrestrial life is reflected in his serviceable weight-bearing lower limbs.

I should like to thank all the people who have helped me with this study. First I must pay tribute to the artist Pamela Arnaud for her line drawings made from somatotype photographs.

Next on the list is Professor J. M. Tanner who inspired us all with his intricate knowledge and who, in fact suggested that I should write this book.

The study spread out over many years–it is, of course a long-term study–hence the data is rather fragmentary. The work was carried out in a Childrens' Home, with an enormous amount of help from Reg Whitehouse (Child Study Centre). I am indebted to him for his expert technical skill coupled with a readiness to help, both in the Childrens' Home, and Child Study Centre.

We have to thank E. R. Bramsby for his work in the early days; Mr. A. C. Rains for reading the script, and for helful criticism; C. Lloyd Roberts for reading the work and supporting it; Harry Asher for permission to use anatomical drawings; Dr. R. Mackeith for reading the script and making practical suggestions; Carole Wilkinson my invaluable secretary, who was full of helpful suggestions; Florence Brush who was very appreciative of the book; and finally Doris Hite for her constant support and readiness to help.

C. A.

CHAPTER 1

1 Relationship of Postural Variations to General Growth Pattern

As the child grows, variations in posture occur; these are not 'defects', they are part of child's method of response to the demands of gravity. Examples of postural variations met with in childhood are as follows:

	Age group	
Bow legs	1–3	(*see* page 73)
Knock Knees	2–6	(*see* page 54)
Kyphosis	Adolescence	(*see* page 26)

These and others are described in the text. An outline of the general growth curve is shown in *Figures* 1.1, 1.2, 1.3 and 1.4, i.e, growth in height and weight in boys and girls, and short accounts of changes in body components and body proportions are given.

Growth in height of girls

In general, the rate of growth in height diminishes rapidly in the first two years and continues to slow down in the pre-school period; there is very little, if any, increase in velocity in the early years of school life; the child gains about the same amount each year, both in height and in weight. Boys and girls on average do not differ very much in size before the age of nine years, though girls are slightly fatter and a little shorter than boys, and boys are, on average, a little more muscular. After the ninth birthday girls begin to grow more rapidly; this increase in rate of growth continues for two or three years. Maximum velocity is reached by the average girl at the age of twelve; the increments then become smaller and after three years there is no further increase in the amount of height gained. Between the ages of sixteen and eighteen, growth in height ceases. It will be noted that the peak in velocity occurs at the age of twelve, about a year before the menarche.

Growth in height of boys

Boys have a similar growth spurt, but it begins two years later than that of the girls, on average at about eleven years old, and is at its

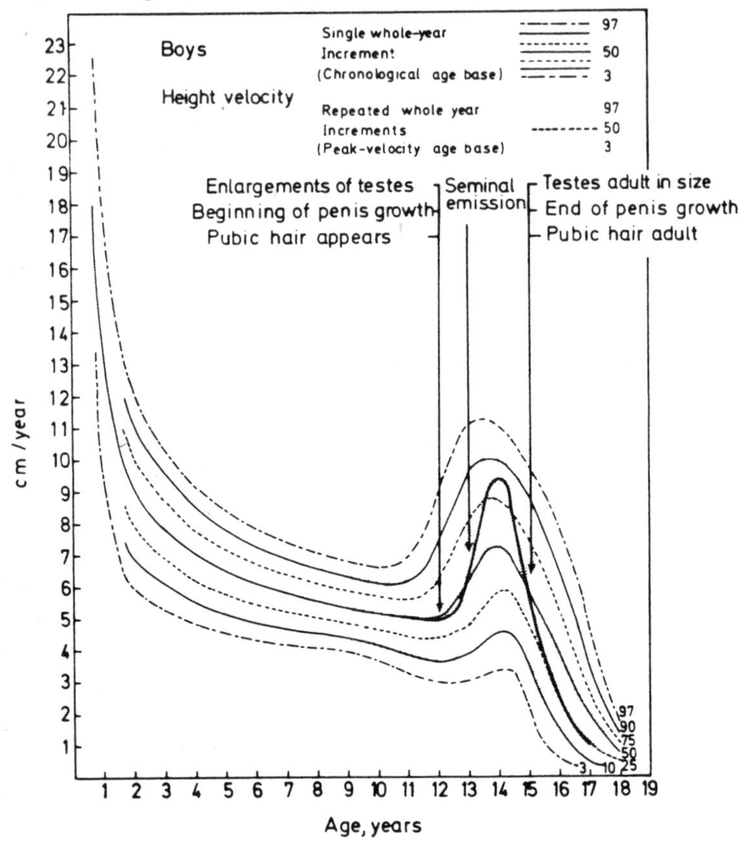

Figure 1.1. Height velocity curve; boys (Tanner & Whitehouse, 1966. Reproduced by permission of the Editor, Archives of Disease in Childhood.)

maximum velocity at about the age of fourteen. After this age increments decrease steadily in size; growth is slowed down, and finally ceases in the early twenties.

Weight in girls and boys

(*Figures* 1.3 and 1.4) In both sexes, there is a marked decrease in the amount of weight gained annually during the first two years of life. During the pre-school years, gain in weight is slow; in the early school years, increases in height and weight are steady. When the height spurt begins, in both sexes, weight increments become larger; the weight spurt, however, is spread over a longer period than the height spurt.

Girls tend to lose weight after adult stature is attained (about seventeen years old); they look slimmer than at fifteen. Boys continue to gain weight up to the age of twenty-one; muscle accounts for much of this increase in weight.

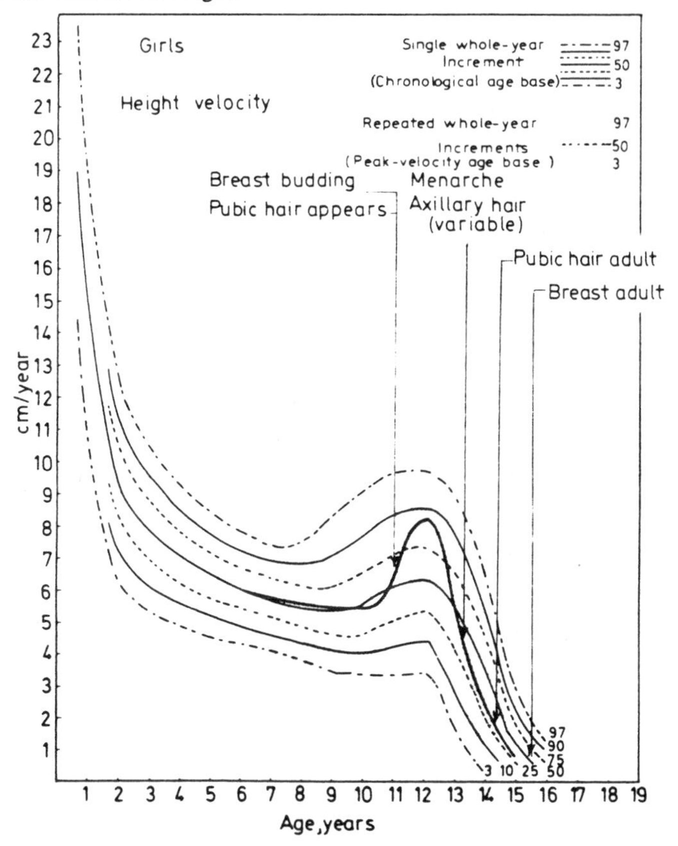

Figure 1.2. Height velocity curve; girls (Tanner and Whitehouse, 1966. Reproduced by permission of the Editor, Archives of Disease in Childhood.)

The age at which the growth spurt begins varies; the range in girls is from eight to thirteen years, and in boys from ten-and-a-half to sixteen years. In a class of eleven-year-old girls, some would not have begun to grow more rapidly; some would have reached their peak in velocity; others might be menstruating and be well past the peak age of growth.

Changes in body proportions

Changes in body proportions are brought about by differential growth rates of skeletal tissue. During infancy growth is most rapid, first in

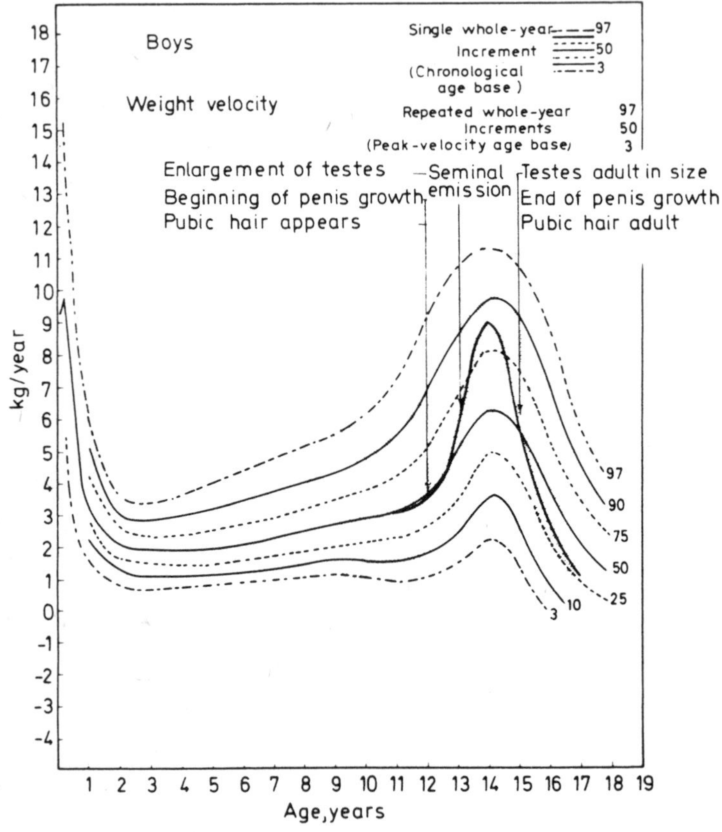

Figure 1.3. Weight velocity curve; boys (Tanner and Whitehouse, 1966. Reproduced by permission of the Editor, Archives of Disease in Childhood.)

the head and later in the trunk. In the second year the legs begin to grow more quickly than the body, and this pattern continues until the onset of the growth spurt at puberty, when in both sexes the trunk grows faster than the limbs.

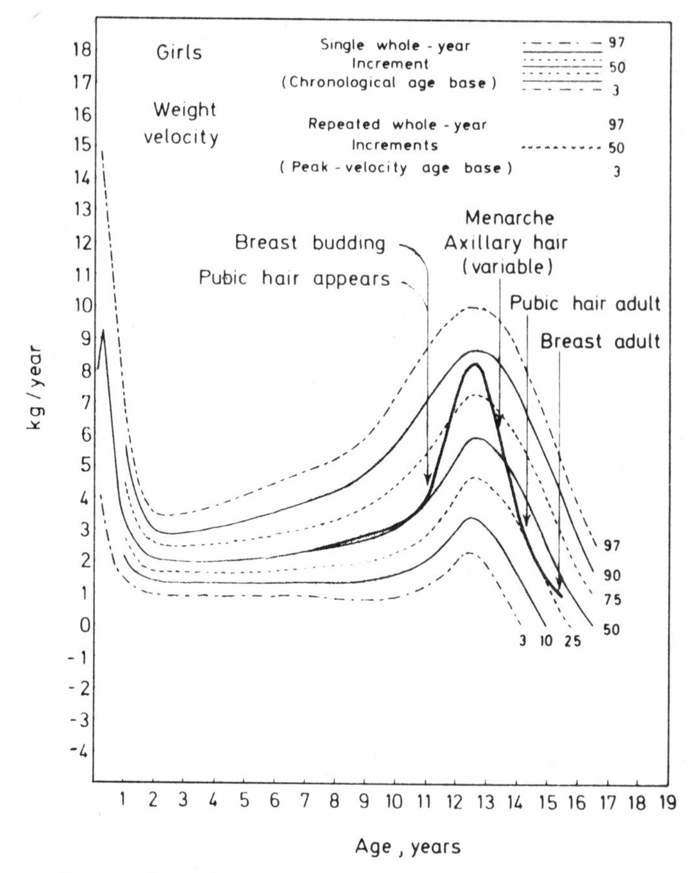

Figure 1.4. Weight velocity curve; girls (Tanner and Whitehouse, 1966. Reproduced by permission of the Editor, Archives of Disease in Childhood.)

As the growth spurt begins about two years later in boys than in girls, it follows that the growth of the lower limbs will continue longer at prepubertal rate in boys, and therefore the legs/trunk ratio will be greater in boys than in girls. For the same reason, late maturing girls may be relatively longer-legged than their earlier-maturing friends. Changes occur in lateral as well as in linear proportions at puberty. The bones of the pelvic and shoulder girdles grow laterally, so that both bi-acromial and bi-iliac diameters are increased.

In boys, bones constituting the shoulder girdle grow more rapidly than those of the pelvic girdle; in girls, the pelvic girdle grows more quickly than the shoulder girdle. Thus the boy becomes broad-shouldered and long-legged, and the girl becomes broad-hipped and has relatively shorter legs. In girls one of the first signs of pelvic growth is a widening of the hips; this is partly due to fat deposits. The pelvis grows slowly and symmetrically before puberty, but, during pubescence, not only is growth more rapid, but re-modelling occurs; the pelvis increases more in width than in depth, and the forepart is widened and rounded (*Figure* 1.5). This period of rapid growth and fundamental change takes eighteen months; it begins soon after the first signs of puberty and finishes soon after menarche.

Changes in body components

Whereas the increase in height depends on the rate of skeletal growth, increase in weight depends on the rate of increase, not only of bone, but of fat, muscle and viscera. The proportions of these components in the body vary with age and sex.

Fat

Fat can be estimated by skinfold measurements in various parts of the body, and by radiological measurements of the calf and thigh. Babies put on fat rapidly in the first nine months of life and then lose it steadily until they are about six years old; girls lose fat less rapidly than boys. From about six onwards, in both sexes, subcutaneous fat begins to increase slowly and steadily in amount.

Fat has its own growth spurt in boys; it starts about a year before the height spurt and lasts about two years. When the growth spurt is well under way loss of fat occurs, but, as soon as the general growth spurt is over, fat again increases in amount, Girls are fatter than boys at all ages after two; they lose fat more slowly between one and six. There is probably a small increase in subcutaneous fat a little before the adolescent spurt, and continuing until the general spurt has ended. It follows that fatness and thinness have not always the same significance. The baby is fat; the pre-school child is thin; this is what we should expect. Some children, however, are always fatter than others; this may be due to the high degree of endomorphy in their make-up, and is probably genetically determined. Obesity in childhood may arise in an endomorph who eats more than he needs; he will become even more obese when the fat spurt begins. On the other hand, mesomorphs may put on fat in middle childhood;

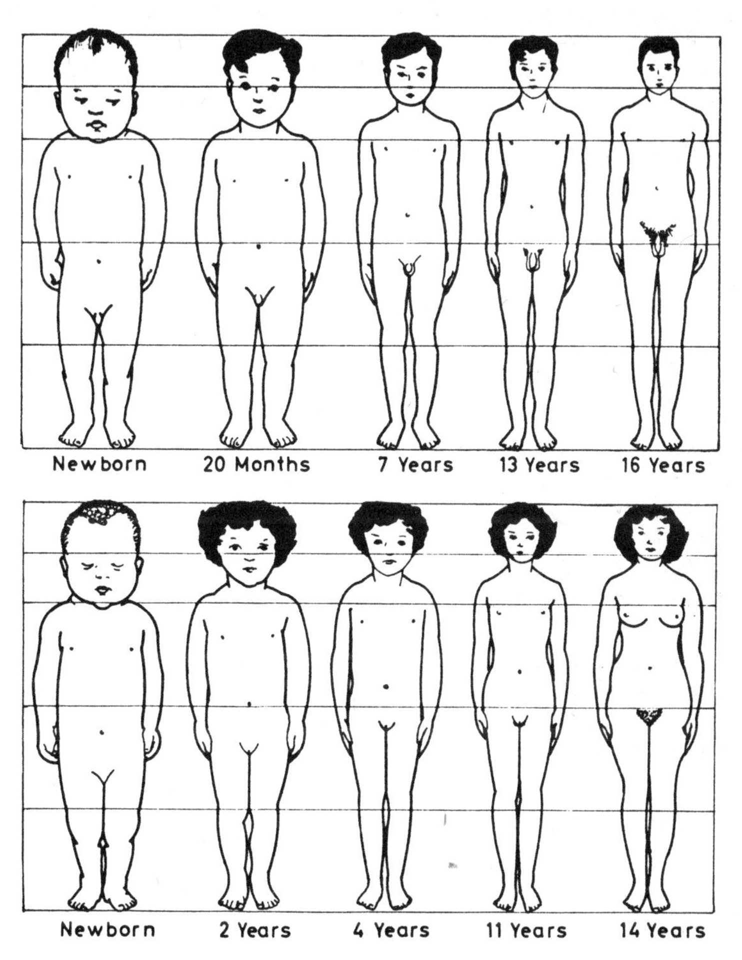

Figure 1.5. Changes in body proportions (Ellis, 1947. Reproduced by permission of the publishers.)

they respond very well to dieting, and rarely acquire the postural patterns (lordosis, knock knee, prominent abdomen) so common in endomorphs.

Muscle

From the middle of prenatal life to early maturity the largest part of the weight increment is made up of muscle. At birth, 25 per cent of body weight is due to muscle; by early adolescence the proportion is about 33 per cent, and by the age of twelve, it may be 43 per cent, Thus the gain in musculature is nearly equal to the gain in all other organs, systems and tissues. The leg muscles grow more rapidly than those of the arms and trunk; the muscles of the head have the smallest increment. At adolescence, strength increases as well as size. Girls appear to arrive at their maximum strength at the time of the menarche, but boys may not achieve their maximum level until growth in height is finished.

Constitution: physique: somatotypes

The general pattern of growth in the median child has been outlined. Children of the same sex and age, however, vary in build; each has a shape which is characteristic of his constitution, and, with certain modifications due to growth, will persist through life. It is generally accepted that body build is genetically controlled, though it cannot become apparent until the genotype has reacted with the environment to produce the phenotype. We can express the kind of body build of an individual by means of the somatotype technique; that is to say, the physical part of constitution is measurable.

Constitution Harrison (1958) defines constitution broadly as the sum total of structural, functional and psychological traits of individuals; it implies a connection between the individual's shape and his character and temperament.

Physique is used to mean bodily structure.

Somatotype is the name given to rating obtained when using Sheldon's 7-point scale to assess proportions of endomorphy, mesomorphy and ectomorphy present.

Classification of body build: somatotypes

Many different methods of classifying body build have been tried; for example in 1925 Kretschmer described pyknic, athletic and asthenic types, and Davenport (1924) described fleshy, medium and slender types. It was suggested than an individual of one type was more liable to certain diseases or to mental disorder than one of another type. Individuals, however, do not fit easily into such pigeon-holes; each may be regarded as a mixture of different components. Sheldon (1940) drew attention to the continuous nature of variation, and names these components endomorphy (roundness), mesomorphy (muscularity) and ectomorphy (linearity). The somatotype can be expressed as three digits, each component being rated on a 7-point scale. Thus an individual high in endomorphy (roundness), low in mesomorphy (muscularity), and low in ectomorphy (linearity) would be rated 711; an example of extreme muscularity would be 171, and of extreme linearity 117.

The somatotype of the adult may be estimated by careful inspection of somatotype photographs and by comparison with standards; Sheldon's original tables were built up from measurements of 1000 nude students.

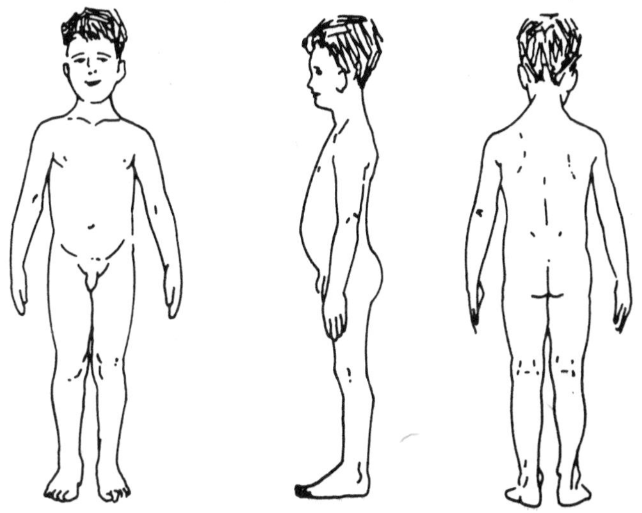

*Figure 1.6. Somatotype drawing of a child with a high degree of endomorphy**

Until recently, somatotyping of children has not been possible because standards have not been available; even without standards, however, the somatotype could often be predicted with some degree of accuracy by experienced observers during early years of school life and it is usually possible to identify the predominant component from the somatotype

* The line drawings were copied in black ink from somatotype photographs. The artist has altered the faces.

drawing. Three illustrations of somatotype line drawings of eight year-old boys are reproduced, and characteristics which might be observed in children high in one of the three basic components are described.

Endomorphy (Figure 1.6)

Petersen (1967) gives the following description of an individual with a high degree of endomorphy:

The endomorphic person is described as rounded and soft, with central concentration of mass; the trunk predominates over the limbs, and the abdomen over the thorax. The proximal segments of the limbs predominate over the distal. The muscles are poorly developed. The contour is smooth, the head is large and round and the face wide. The neck is short, the shoulders high and rounded. The trunk is long, the back straight and the chest wide. The abdomen is large and the waistline high. The limbs are short and tapering, the skin is soft and velvety, the genitalia are hypoplastic.

Mesomorphy (Figure 1.7)

Some of the characteristics of mesomorphy described below can be seen in the somatotype drawing.

The body is sturdy, hard and firm; the bones are large and heavy, the muscles well developed, massive and prominent. The transverse diameters are as large as those of the endomorph but the antero-posterior diameters are not so large. The muscular thorax predominates over the abdomen; the jaws are square, the shoulders broad, the clavicles robust, the knees are strongly built and the ankles are bony. The face is large with massive cheekbones, the trunk long, the thorax wide, the buttocks laterally dimpled, and the abdominal muscles are visible. In fact, children high in mesormorphy look tough.

Ectomorphy (Figure 1.8)

Linearity, flatness of chest and a certain delicateness can be seen in the child high in ectomorphy. Visceral and somatic structures are poorly developed; the face is smaller than the cranium; dolicocephaly is

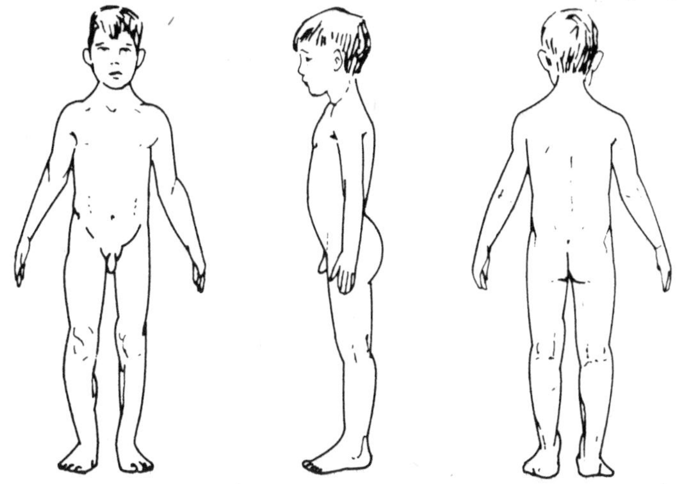

Figure 1.7. Somatotype drawing of a child with a high degree of mesomorphy

common and the face is triangular in shape. The neck is long and slender, the shoulders droop, and the neck may poke forwards.

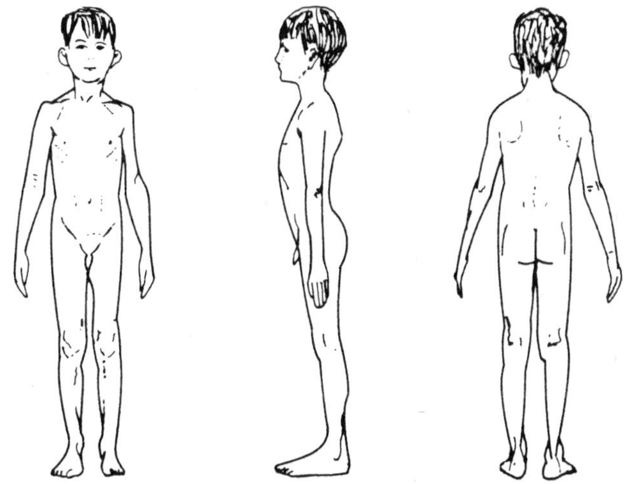

Figure 1.8. Somatotype drawing of a child with a high degree of ectomorphy

The trunk is short, and the extremities are long; the arms and legs are long, especially the distal segments.

REFERENCES

Davenport, C. B. (1924). *Body build and its Inheritance,* Washington, D. C. Carnegie Institute Publications No. 329

Ellis, R. W. (Ed) (1947). Now Mitchell, R. G. (1970). *Child Life and Health,* London and Edinburgh: Churchill Livingstone

Harrison, R. J. (1958). *Man: the pecular animal,* London: Penquin Books

Kretschmer, E. (1925). *Physique and Character,* New York: Harcourt Brace & Co.

Peterson, G. (1967). *Atlas for Somatotyping Children,* Netherlands: Van Gorcum Ltd

Sheldon, W. H. (1940). *The Varieties of Human Physique,* New York: Harper

Tanner, J. M. and Whitehouse, R. H. (1966). 'Standards for skeletal maturity based on a study of 3000 British children' *Archs Dis. Childh.* **41,** 454, 613

Additional reading:

Bayer, L. M. and Bayley, N. (1959). *Growth Diagnosis,* Chicago: University of Chicago Press

Greulich, W. W. and Thomas, H. (1944). 'Growth and development of pelves of individual girls, before, during and after puberty.' *Yale. J. Biol. Med.,* **17**, 91

Marshall, W. A. and Tanner, J. M. (1969). 'Variations in pattern of pubertal changes in girls'. *Archs. Dis. Childh.,* **44**, 235

Reynolds, E. L. and Wines, J. V. (1948). 'Individual differences in physical changes associated with adolescence in girls.' *Am. J. Dis. Child.* **75**, 329

Stuart, H. C. (1946). 'Normal growth and development during adolescence.' *New Engl. J. Med.* **234**, 666, 693 and 732

Tanner, J. M. (1962). *Growth at Adolescence,* 2nd edn. Oxford: Blackwell

CHAPTER 2

2 Study of Postural Patterns in Childhood

Introduction

Postural patterns in childhood vary with age, sex, stage of development and body type. In this study variations in normal children are discussed and the incidence of so-called postural defects are reviewed.

Method of study

The data used were obtained from clinical examinations of a group of children.

Note on subjects of study

Number of children involved in growth study		600
Date of starting and finishing		1950: 1970
Age range	Entry	1–6 years
	Leaving	2–20 years
Average stay		4 years
Range		1 year to 18

These children were examined anthropometrically, as part of a growth study; they attended for examination twice yearly.

Obviously here was an opportunity to make *clinical* examinations of normal children, and to find out how patterns developed and changed during childhood and adolescence. The research team kindly allowed me to carry out my own examinations on their subjects.

The children lived together in a Children's Home run by a voluntary society. In order to create a home-like atmosphere the children were housed in small family groups each with a housemother in charge. The Home is situated in an open space, on high ground; the houses (converted to flats) surround a large stretch of grass, on which are several fine oak trees. Material conditions are good; the diet is excellent; children are treated as individuals, and every effort is made to compensate them for the deprivation of normal home life. The age of entry varies. Most children, on admission to the Home, are already attending school, while others remain until school-leaving age or later. Arrangements are made, when the time comes, for training, or for placement in suitable jobs; some of the children return to their parents, but many find lodgings near the Home. Whenever possible, the children are followed up by the Research Team and many of them return annually for examination, often in due course bringing with them their own offspring.

Procedure

Each child attends for examination every six months. The medical records kept by the medical practitioner and the nurse attached to the Home are available for inspection; illnesses which may have occurred since the last visit are noted. A brief clinical examination is carried out on each child. This is followed by a detailed postural survey which includes taking a record of the footprint with Scholl's pedograph and measurement of pelvic tilt with Wiles' inclinometer. Each child is inspected in four positions, lateral, posterior, anterior and long-sitting. The observations on posture are recorded on a standard chart. (*Figure* 2.4).

Lateral (*Figure* 2.1)

The child is seen in profile; he is asked to stand easily and naturally. The position of the head is recorded, also that of the shoulders; the contour of the abdomen is inspected, and the presence or absence of kyphosis or lordosis is noted.

Posterior (*Figure* 2.2)

The child is asked to turn his back to the examiner; the position of the scapulae, and the presence or absence of lateral curvature is noted, and the landmark used for measuring the pelvic tilt is marked with skin pencil or biro (level of posterior superior spines). The feet are inspected for 'valgus heel' (*see* page 81).

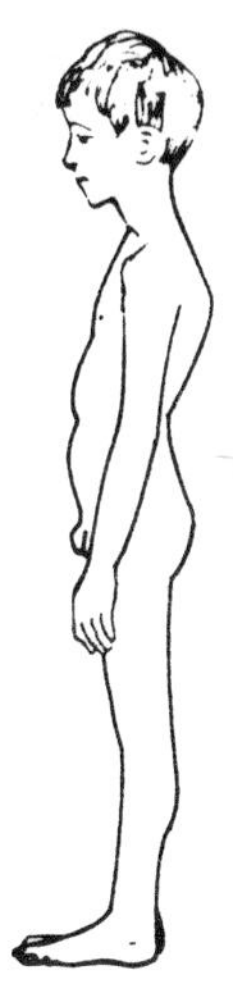

Figure 2.1. Eight-year-old boy, lateral view

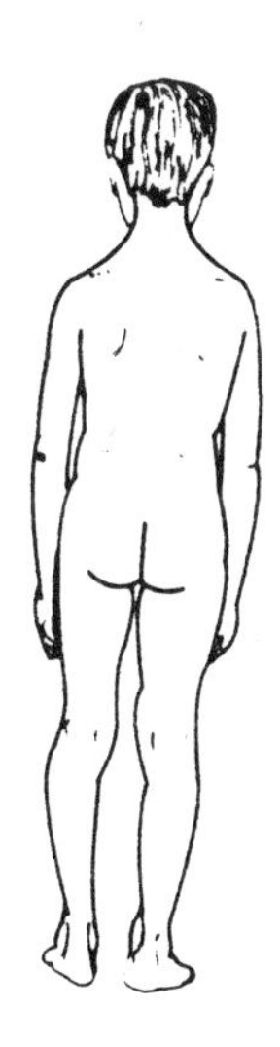

Figure 2.2. Eight-year-old boy, posterior view

Anterior (*Figure* 2.3)

The child is asked to face the examiner. The general posture is appraised; the direction of the patella, taken in conjunction with the direction of the forefoot is noted. The feet are inspected for hallux valgus and metatarsus varus.

Figure 2.3. Eight-year-old boy, anterior view

Long-sitting position

The child sits on the floor, with his legs stretched out in front of him. He is asked to touch his toes without bending his knees and the presence or absence of tight hamstrings is noted. The calf muscles are tested for tightness by dorsiflexing the foot at the ankle joint; the degree of knock knee (if present) is estimated.

Erect position

The child is asked to stand up, and the pelvic tilt is measured with the inclinometer. At some time during this examination the footprints are recorded; this procedure interests and amuses the children; it is often carried out before the examination as it helps to establish rapport.

Posture during exercise

The child is asked to exercise; he is encouraged to run round the room, and also to jump up and down, as though skipping. His postural pattern is noted before, during and after exercise.

POSTURE CHART

EXAM. DATE ______________ BIRTHDATE ______________ NAME ______________________

Item	Column		Code		Code	
STUDY	1					
RACE	3	E	1	N/E	2	
SEX	4	BOY	0	GIRL	1	
EXAM. DECIMAL	5	☐☐☐☐☐				
BIRTH DECIMAL	10	☐☐☐☐☐				
AGE DECIMAL	15	☐☐☐☐				
NUMBER	19	☐☐☐				
NECK	22	STRAIGHT	0	FORWARD		1 2 3
SCAPULAE	23	FLAT	0	WINGED		1 2 3
	24	HIGH	1	MEDIUM	2	LOW 3
SHOULDERS	25	NOT FORWARD	0	FORWARD		1 2 3
SPINE	26	NORMAL	0	KYPHOSIS		1 2 3
	27		0	SCOLIOSIS		1 2 3
	28		0	LORDOSIS		1 2 3
ABDOMEN	29	NOT PROMINENT	0	PROMINENT		1
	30			PROMINENT ABOVE		0 1 2
				PROMINENT BELOW		3 4 5
				PROMINENT SMOOTH		6 7 8
KNOCK KNEES	31	1" or less	0	1" - 2"	1	2" or more 2
BANDY LEGS	32	1" or less	0	1" - 2"	1	2" or more 2
INTERNAL CONDYLES	33	NOT LARGE	0	LARGE	1	2 3
PATELLAE	34	IN	1	STRAIGHT	2	OUT 3
TIGHT HAMSTRINGS	35	ABSENT	0	PRESENT	1	2 3
TIGHT CALF MUSCLES	36	ABSENT	0	PRESENT	1	2 3
LONG ARCHES	37	FLAT 1 2 3		MEDIUM	4	HIGH 5 6 7
TOES GAP	38	ABSENT	0	PRESENT	1	2 3
HALLUX VALGUS	39	ABSENT	0	PRESENT	1	2 3
POSITION OF FEET	40	IN	0	STRAIGHT	2	OUT 3
VALGUS ANKLES	41	ABSENT	0	PRESENT	1	2 3
OVERALL POSTURE	42	EXCELLENT	1	GOOD	2	FAIR 3 BAD 4
PELVIC TILT	43	☐☐				
PHYSICAL STATUS	45	THIN	1	AVERAGE	2	PLUMP 3 OBESE 4
TYPE	46	STOCKY	1	INTERMEDIATE	2	SLENDER 3
MENARCHE	47	NOT YET	0	YES	1	
"	48-51	MONTH/YEAR	☐☐☐☐			

9 in any column denotes NOT RATED

Figure 2.4. Posture card

CHAPTER 3

3 General Postural Patterns of Childhood

Intense general mobility is characteristic of the child's behaviour throughout the early years of school life. Scrambling, jumping, running, climbing and above all wriggling are just a few of the activities which occupy his leisure hours. If on examination he is asked to stand still, small, extra movements of hands and feet, clawing of the feet, and flexion of the toes occur. The angle of the inclination of the pelvis is altered constantly, the neck pokes forwards and the head retracts. The scapulae may appear to be mobile one minute, and flat on the chest a moment later. As children grow older, they become more quiescent; they tend to return to positions distinctive to themselves after bouts of activity.

It should be quite clear however that there is no question of 'ontogeny repeats phylogeny'. Ontogeny refers to the development of race; the child *in utero* does not recapitulate the stages through which the race has passed.

In order to understand, not only the anatomy, but the behaviour patterns which can be observed at different stages, these evolutionary factors must be appreciated and no apology is needed for including in this chapter a short account of development of the upright posture. (*see also* lead article in *Lancet,* 1964).

EVOLUTION OF THE UPRIGHT POSTURE

The human child as we know him today is the product of millions of years of natural selection; the process whereby species showing variation favourable to environment tend to be preserved, and those showing unfavourable variations tend to be destroyed. Many of the phases of skeletal development which can be observed in childhood today have their origins in pre-historic times and many aspects of behaviour are reminiscent of life in the trees.

Man is a member of the primate family, a family which includes lemurs, monkeys, tarsiers, apes and man (*see* Le Gros Clark, 1962). The first mammals probably appeared on land 90–50 million years ago, after the collapse of the reptiles; they were small insect-eating, shrew-like creatures, who, on account of their mobile limbs, could support themselves in water and drag themselves about on land (*Figure 3.1*). Pronation, developed during aquatic life, was natural to them; in course of time they became adapted to clambering over objects on the ground and then to tree-climbing. Gradually hind limbs were adapted for weight bearing, hands for grasping food and for holding interesting objects, the better to examine them and satisfy curiosity. Five digits were developed on each of the four limbs; fingers and toes became mobile; the clavicle appeared, and was used as a strut when arms were moved sidways.

Skill and agility were necessary to these tree-dwellers. Gradually, by selection pressures, eyes were moved further forward, making three dimensional vision possible; the brain increased steadily in size, as vision improved. These early, nimble, tree creatures led busy active lives leaping through the branches in search of food, which consisted mainly of fruit, nuts and leaves, with occasional grubs or insects. A million years ago they probably bore some resemblances to the monkeys we know today. Many monkeys still prefer the trees, but the ape-like creatures who were man's ancestors probably left the forest fifteen million years ago, to lead a new life on the ground.

Why did our ancestors leave the trees? Probably a time came when, owing to climatic conditions, the face of the earth changed, dense forests no longer flourished; in their place small woods appeared dotted about the landscape, separated from each other by stretches of scrub and savannah. Terrestrial life must have been hard going for these early tree-dwellers, accustomed as they were to a mainly vegetarian diet. They found themselves in direct competition with four-footed animals, fleet of foot and adapted over millions of years to catching prey and meat-eating. In course of time, however, our ancestors' speed improved as selection pressures favoured a more upright posture; they began to hunt for food, and later to eat and digest meat.

Adoption of the upright posture was associated with freeing of the upper limbs from locomotion, enlarging the field of vision (with increase in size of the brain) and with tool-making. It is possible that tool-making developed before walking and standing had been perfected, and may have been one of the evolutionary factors responsible for the attainment of the upright posture.

Hamerton (1962) suggests that, in some of the common ape ancestors, probably those living on the edge of the forest, some of

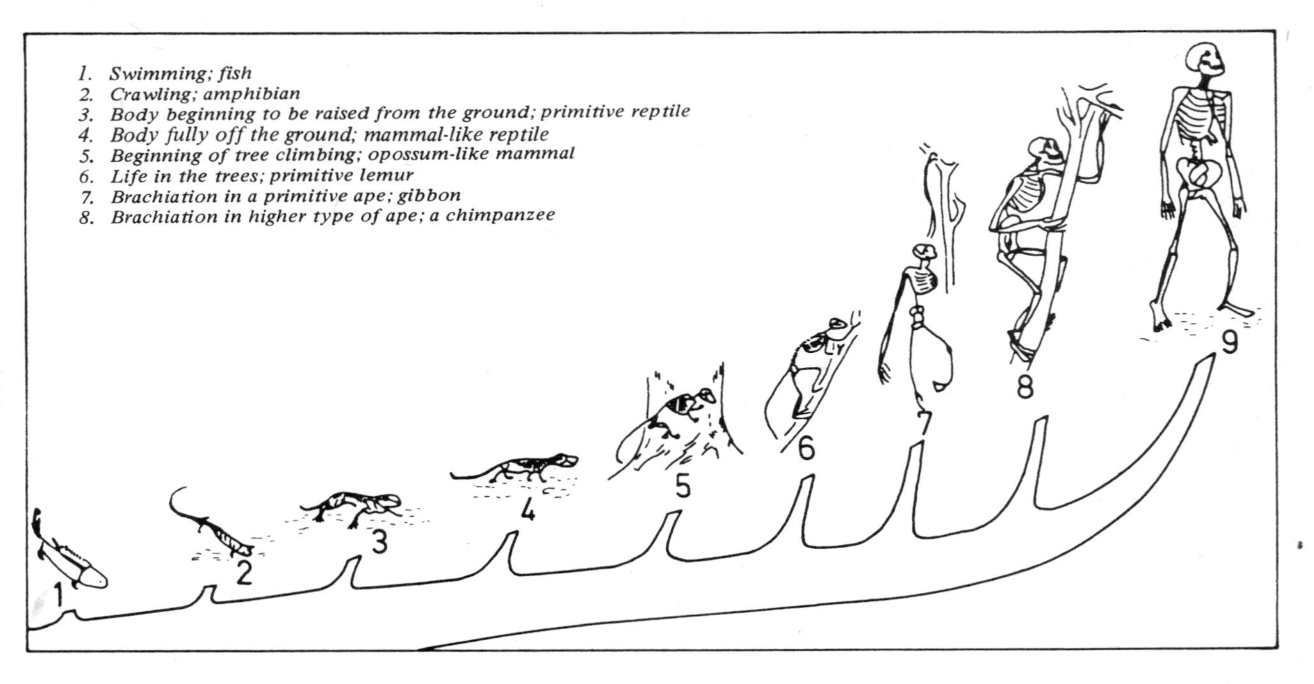

Figure 3.1. The evolution of man and his upright posture (From Wells, Huxley and Wells, 1946)

the translocations of the chromosomes favoured bipedalism rather than brachiation, and that the ape-like creatures possessing these mutations went out into the open country to start life in the savannah.

Washburn (1960a) points out that the critical primary adaptation to bipedalism was through shortening of the ilium, which had also to be bent back for obstetric reasons (*see also* Washburn 1960,b). In some of our ancestors, selection pressures caused the gluteus maximus to move far enough behind the hip joint for it to become an extensor rather than an abductor, thus making possible the basis of upright posture and gait. This he thought was possibly the essential change brought about by the initial chromosome translocation. Harrison suggests that, paradoxically, the upright posture may have been developed in an attempt of our ancestors, through selection, to retain tree life; the erect posture would enable the open space between scrub and savannah to be covered more rapidly and effectively.

ANATOMY OF THE SPINE AND ITS CONNECTIONS

Vertebral column (*see* page 22)

(1) *Curves of the Spine: Primary and Secondary*

Cervical Curve; convex forwards	Secondary; 2–3 months after birth	Compensatory: develop as intervertebral discs become wedge-shaped
Thoracic curve; concave forwards	Primary; present at birth	Due to variations in depth of bodies of vertebrae
Lumbar curve; convex forwards	Secondary; appears when child learning to stand erect	Compensatory due to variation in depth of intervertebral discs
Sacro-coccygeal curve; concave forwards		

The spine is composed of 33–34 vertebrae; it can give flexibility or rigidity as occasion requires. Each vertebra is constructed on the same plan, and consists of a body and an arch; the bodies are separated from each other by the intervertebral discs, which are made of tough fibrocartilage; the vertebral arches unite to form a canal, which gives protection to the spinal cord.

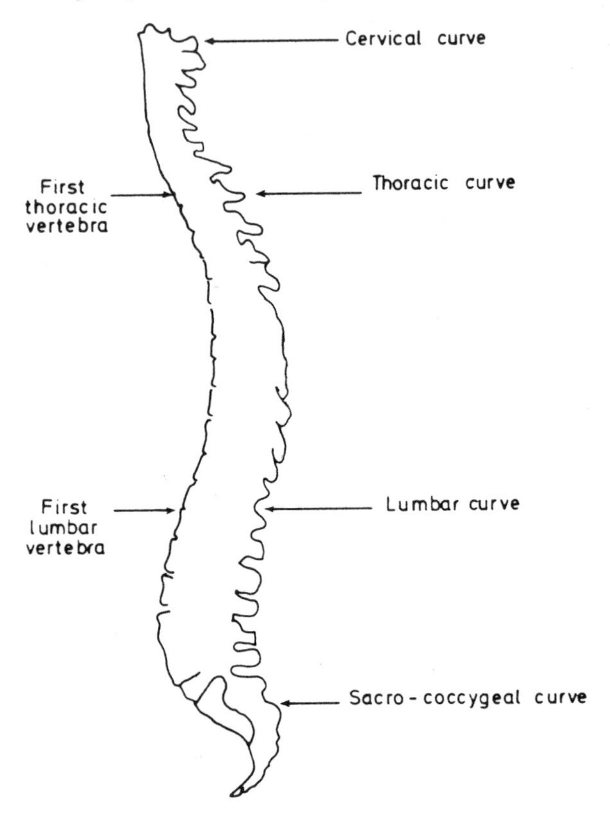

Figure 3.2. Human vertebral column

When our ancestors began to stand erect, certain changes were necessary. The head had to be balanced on the upper end of the spine, so that the eyes could face forwards; the head and trunk had to be balanced over the lower limbs by means of the pelvic girdle; the whole body had to be balanced over the space occupied by the feet when the soles were planted firmly on the ground (centre of gravity (*see* Chapter 5). These manoeuvres were made possible by the development of four antero-posterior curves (*see Figure 3.2*).

The fetus lies comfortably in a position of flexion in his uterine bath; his spine is disposed in the shape of one long kyphotic curve, concave forwards (which includes cervical, dorsal and lumbar vertebrae) and one shorter sacro-coccygeal curve. (*Figure* 3.3).

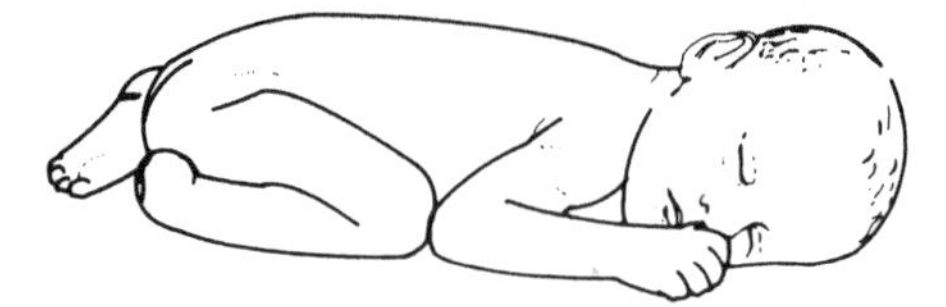

Figure 3.3. Premature baby, showing long kyphotic curve (Illingworth, 1972. Reproduced by permission of Churchill Livingstone.)

When, in the early months of postnatal life, the child extends his head, a small compensatory *lordotic curve* (*see Figure* 3.4), convex forwards, appears in the cervical region; when he begins to

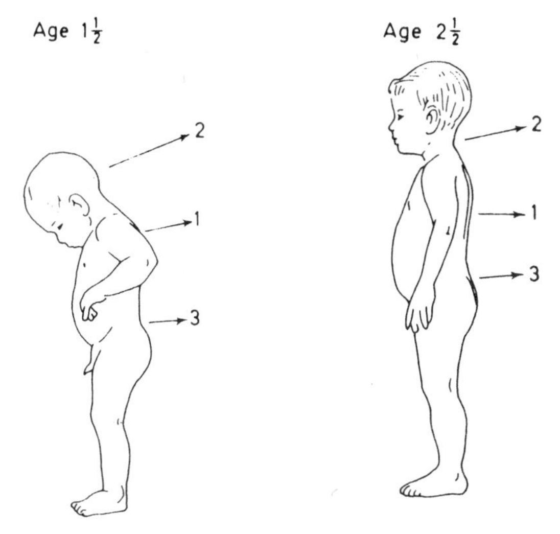

Figure 3.4. Curves of the spine

1. Thoracic, primary, convex (kyphotic)
2. Cervical, secondary, concave (lordotic)
3. Lumbar, secondary, concave (lordotic)

sit up a secondary lordotic curve, convex forwards appears in the lumbar region. The secondary convex curves depend on differences in thickness of the intervertebral *discs* which become wedge-shaped to allow for necessary adaptation; the primary curves, on the other hand, depend on differences in height between the anterior and posterior aspects of the *bodies* (*Figure* 3.4).

During the first two years of life the lumbar vertebrae grow rapidly, with consequent lengthening of the lumbar region and also of the loins. The lumbar development is probably associated with walking on two feet; longer muscles make walking easier and more efficient.

(2) Connections of the vertebral column

The vertebral column articulates with the occiput, the ribs, the sacral vertebrae and the hip bone; there is no direct articulation between the vertical column and the shoulder girdle. Each vertebra articulates with one vertebra above and one below by the superior and inferior articular processes; the bodies are separated from each other by fibrovascular cartilage (the discs). Strong ligaments connect the bodies anteriorly and the spines posteriorly; the curves of the spine are adjusted to the exigencies of the moment by means of a complex system of muscles; in particular some muscle fibres pass from the transverse processes of the lower vertebrae to the spines of those above (multifidus and semispinalis) so that local flexions and extensions are possible.

(3) Movements of the vertebral column

Flexion is the most extensive of all movements of the vertebral column; it is most free in the lumbar region, whereas extension is most free in the cervical region. Lateral movement is greater in the lumbar spine than in the thoracic. The rib cage limits thoracic lateral tilt. The first two cervical vertebrae, the atlas and the axis, differ from the remaining five; the atlas has no body and the space where one might expect to see a body is occupied by the adontoid peg of the axis. The head inclines backwards and forwards, in flexion and extension at the atlanto-occipital joint as in nodding; the only movement which can occur at the atlanto-axial joint is rotation. Muscles which run from transverse processes of vertebrae to spines higher up in the vertebral column (semispinalis cervicis and multifidus) draw the neck backwards, but the *head* is tilted back by muscles running from the articular processes of the cervical vertebrae, and the transverse processes of the lower cervical vertebrae, and the transverse processes of the upper thoracic vertebrae to the occiput (semispinalis capitis and obliquae capitis superioris). The *head* is *flexed* on the spine by muscles arising from the transverse processes of the lower cervical vertebrae, which are inserted into the basilar part of the occipital bone. It is important to distinguish between head and neck movements when evaluating posture.

The shoulder girdle and attachments

The scapulae, the clavicles and the manubrium of the sternum constitute the shoulder girdle in man. There is no direct articulation between the upper limb and the trunk; the only bony connection is the clavicle, which articulates with the sternum by its inner end and with the acromion of the scapula by its outer end. This arrangement results from our ancestors' way of life; a quadruped who suspended himself by means of his fore limbs (brachiated) had to be versatile in his movements; over the years, circumduction was favoured by selection pressures, the scapula migrated posteriorly and medially, and the insertion of pectoralis minor moved from the humerus to the coracoid. Clavicles were of no use to the mammals who descended early from the trees (for example, the ancestors of the cat and dog families); their upper limbs needed to approach the midline and their clavicles became rudimentary or disappeared altogether. Man's arboreal ancestors, on the other hand, needed a well developed clavicle to act as a fulcrum in allowing movement from side to side. Other features of importance which originate from the period when man lived in the trees are opposability of the thumb and the appearance of five digits (fingers or toes) in lower limbs; both these modifications are necessary for grasping or prehension.

Pelvic girdle: lower limb complex

It is impossible to discuss mechanisms and variations in the pelvic girdle without referring to its origin. Krogman (1951) says that, as a piece of machinery, the human frame is such a 'hodgepodge' and makeshift that it is surprising that we get along as well as we do. In all evolutionary time, the most profound changes occurred when man's ancestor changed from a four-legged to a two-legged animal. If we regard a four-footed animal as a bridge, it will be apparent that, when we up-end the bridge, a tremendous mechanical imbalance occurs; the backbone has to accommodate itself to new weight-bearing vertical stresses. This was accomplished in the case of man by changing a single spinal curve to an S-shaped one. When man began to stand on two legs an extra burden was added to the pelvis, which now had to bear the weight of the body. In standing erect man tilted the whole structure upwards, so that the pelvis was at an angle with the backbone instead of parallel to it. To increase the effectiveness of the lower abdominal wall three muscular layers have evolved which criss-cross to give support. By adjustment of the bony pelvis the centre of gravity has been shifted to the level of a transverse line running through the centre of the hip socket. By this means the weight of the trunk upon the pelvis is efficiently distributed on the two legs.

The anatomy of the flexors and extensors of the hip joint is considered later under the heading of 'Kyphosis'.

POSTURAL VARIATIONS

Variations relating to the vertebral column

(1) Kyphosis and tight hamstrings

Kyphosis may be defined as a curvature of the spine, convex backwards. While it is true that kyposis can occur as a pathological sequel to tuberculosis of the spine, and in conditions such as Scheuermann's disease, it is found in a minor degree as a normal occurrence in infancy, childhood, and adolescence, and is reminiscent of the days before our ancestors assumed their upright posture. As described above (page 23), the fetal spine assumes one long kyphotic curve from the first cervical to the fifth lumbar vertebra; the two secondary compensatory lordotic curves appear in postnatal life and shorten the span of the kyphotic curve.

Three grades of kyphosis are listed on the posture card, mild, medium and severe. Those in which there was only a slight exaggeration of the normal curve were graded as mild; there were no 'severe' cases in study. The mild and medium cases were considered together. In boys, no definite pattern emerged; kyphosis occurred more often after the age of eleven than before, and reached peak incidence near the age of sixteen; thereafter it was rarely seen. In girls, mild cases occurred sporadically throughout school life but there seemed to be little if any relation to the stage of development at which the child had arrived. The body type seemed to be of some importance; ectomorphic children, especially girls, were liable to show exaggeration of the normal kyphotic curve at any age.

Pathological kyphosis This condition, first described by Scheuermann in 1920, is seen most often in boys between the ages of nine and eighteen, during the period of active growth; the changes usually occur in the lower thoracic or lumbar vertebrae and this serves to differentiate it from normal postural kyphosis. It manifests itself as a slowly developing long curve not completely correctable; the x-ray shows an irregularity of growth of the epiphyseal plates. It is a self-limiting disease which tends to diminish in about eighteen months, and to stop entirely when growth of spine is complete. It occurs at the time when epiphyseal growth of vertebrae is at a maximum, and the spine is therefore at a disadvantage if exposed to extra strain such as trauma, extra body-weight or tight hamstrings.

Tight hamstrings Lambrinudi drew attention to this condition in 1934. All children in the study were tested six-monthly for tight hamstrings. They were asked to sit on the floor with their legs stretched out in front of them, and to touch their toes without flexing the knees. Three grades were listed on posture card: 0 (absent), 1 (child just failed to touch toes), 2 (two or more than two inches (2.5cm) distance between finger tips and toes).

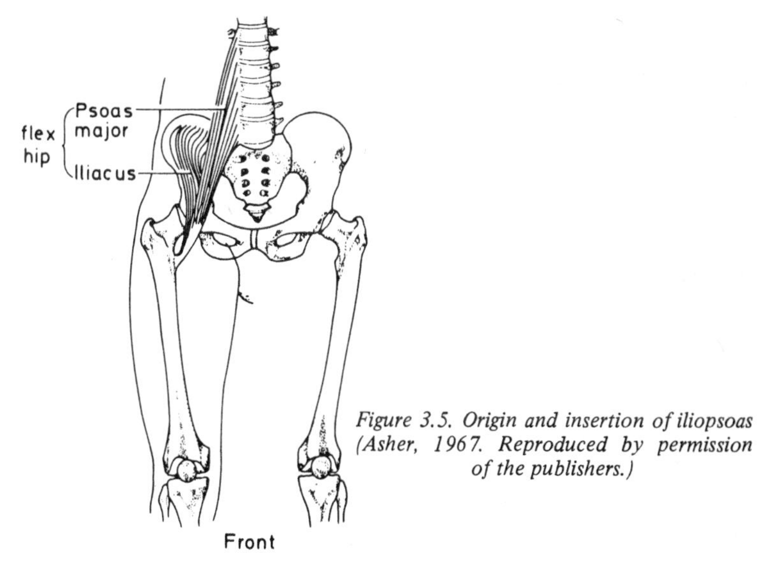

Figure 3.5. Origin and insertion of iliopsoas (Asher, 1967. Reproduced by permission of the publishers.)

Iliopsoas and the hamstrings (Figures 3.5, 3.6) It has been suggested that if the hamstrings are overstretched excessive spinal movements may occur during toe-touching and flexion exercises; growing parts of the vertebrae may be damaged with development of dorsal osteochondritis. Such changes may be irreversible and may become the site of osteo-arthritis in later life. For this reason it is worth reviewing briefly the anatomy of the region.

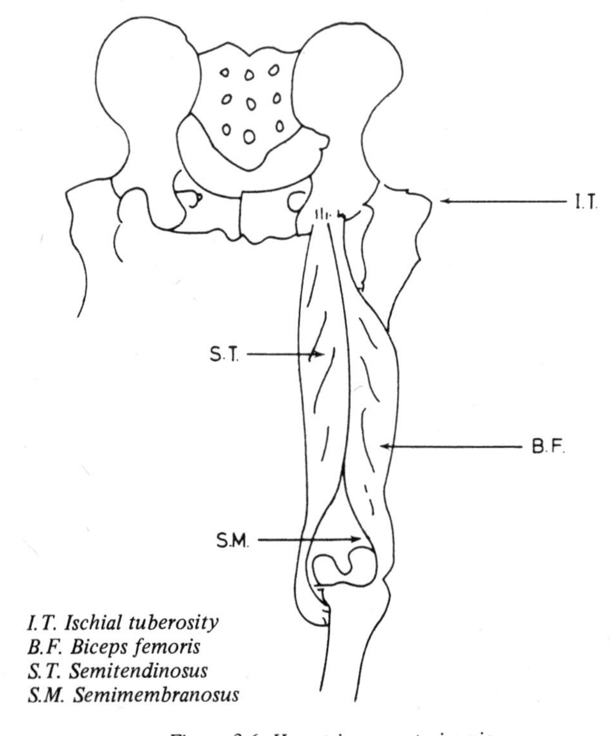

Figure 3.6. Hamstrings, posterior view

Psoas major arises from the transverse processes of all the lumbar vertebrae, the bodies of the twelth thoracic and the first three lumbar vertebrae and from the intervertebral discs and tendinous arches bridging the constricted portions of the bodies of the lumbar vertebrae.

The *iliacus* arises from the upper two-thirds of the iliac fossa, the iliac crest, and the lateral mass of the sacrum; its fibres are inserted into the tendon of the psoas major. The iliopsoas is inserted into the lesser trochanter of the femur.

The *iliopsoas* flexes the thigh on the pelvis; it helps to bend the trunk forwards, and raises the trunk from the recumbent to the sitting position.

Three muscles are included in the hamstring group: semi-membranosus, semitendinosus, and biceps femoris (*Figure* 3.6). All of them arise from or near the ischial tuberosity; the biceps is inserted into the upper end of the fibula, the other two are inserted into the medial aspect of the upper end of the tibia. Acting from above, the hamstrings flex the knee: from below they draw the trunk back or raise it from a stooping position (*Figure* 3.6).

During ordinary sitting the hip is flexed and the knee is slightly flexed; the iliopsoas is responsible for flexion of the hip, and it also enables the child to touch his toes with knees slightly flexed. If the child tries to touch his toes when standing erect with knees extended he may be unable to reach the ground without bending his knees. If he persists in his efforts, the iliopsoas may continue to contract against the resistance of the hamstrings, and it is possible that, over a period of time damage may be done to the origins of iliopsoas (*Figure* 3.5).

In some children, ventral flexion of the thoracolumbar spine is limited for no apparent reason. There may be some fundamental defect; the lumbar spine may be set at an abnormal angle on the pelvis, increasing the lumbar lordosis. Whatever the cause, this 'abnormal' posture may be the 'natural' one for some children, and attempts should not be made to persist with ventral flexion exercises. This is of some practical importance; at the present time, specialists in physical education are aware of this danger. It is possible, however, that in the past damage to lower thoracic spine may sometimes have been caused by injudicious toe-touching.

Incidence of tight hamstrings in childhood From a study of the longitudinal data it appears that it is unusual for tight hamstrings to present themselves before the age of six; they are usually noticed first between the ages of six and eight. The hamstrings do not 'loosen' appreciably with age; in fact the 'tightness' usually persists throughout adolescence to adult life.

Summary In some children limitation of flexion at hip joint occurs. This may be associated with tight (or short) hamstrings, or there may be a fundamental defect in the bony pelvis which leads to increased lumbar curve and limitation of ventroflexion. In itself, limitation of flexion at hip joint is of very little importance. Attempts to lengthen the hamstrings by vigorous flexion exercises, however may lead to damage to bodies of vertebrae, with subsequent wedging and possibly permanent damage. Children who cannot touch their toes should be discouraged from trying to do so.

(2) *Scoliosis*

This was an uncommon finding in the study series. No cases were seen in children under eight; in general, the incidence of scoliosis was higher in boys than in girls, and the age of onset was earlier in boys. There seemed to be no definite relationship between scoliosis and the adolescent spurt. It was commoner in ectomorphs than in mesomorphs. Postural scoliosis manifests itself as a single or total curve, usually convex to the left; it disappears on flexion of the spine, and there is no rotation of the vertebrae. It can be compensatory to deformity of the hip, or to shortening of one leg (Two cases of this kind were seen in the study).

If a child with postural scoliosis is x-rayed, a single lateral curve is seen when the child is standing upright but no curve is seen when he is lying supine. The fact that the spines retain their normal midline position demonstrates that there is no rotation and that the condition is postural in origin. This curvature is of no clinical importance; when it appears faulty postural habits should be corrected, and general exercises may be of benefit. We now know that postural scoliosis is *not* a precursor of structural scoliosis; the one never changes into the other (James, 1951).

(3) *Lordosis*

Lordosis is the name given to the exaggeration of the concave curve which normally exists in the lumbar region. The curve can be increased in concavity by extension of the lumbar spine which causes the child to 'lean back' from the lumbosacral joint; or it can be increased still further by tilting the pelvis forwards; this is effected by flexors of the hip, in particular, the iliopsoas, rectus femoris, and anterior adductors. It can be decreased by contraction of the extensors of the hip, glutei., hamstrings and posterior adductors.

Thus, lordosis, or increased lumbar concavity may be due to extension of the spine; if the pelvis is also tilted, higher degrees of lordosis may occur.

Examples of lordosis with or without increased pelvic inclination may be seen on page 34.

Variations relating to the shoulder girdle

Round shoulders

This is an omnibus term, used to describe children who exhibit all or some of the following features: forward shoulders, poking head, poking neck, kyphosis and mobile scapulae. Children commonly appear round shouldered during adolescence. After the period of extreme mobility during exercise is over the child may assume a round shouldered posture during his position of rest afterwards. In its most marked degree, the neck is poked far forward with head extended (if standing), or flexed following the line of the neck (if sitting at a desk). Shoulders droop, with the acromion directed forwards, and the axis of the upper arm directed backwards; the scapula is slung low, with vertebral angle and vertebral border raised, and possibly directed forwards towards the ribs. Kyphosis may be present. The forward poke of the head sometimes appears to be part of the balancing mechanism (*Figure* 3.7).

Figure 3.7. Girl aged thirteen, round shouldered

A few notes on the individual components of the syndrome follow.

(1) *Forward shoulders* Three different degrees of 'forward shoulders' were listed on posture cards 1, 2 and 3. In most of the primary school children the tip of the acromion process appeared to be pointing forward, with the axis of arm directed backwards; in the older group, especially during the adolescent spurt it was not uncommon for children to be rated 2, which was often seen as position of rest. Thus the forward position was frequently part of the body image

(2) *Mobile scapulae* Three degrees of mobile scapulae were noted; grading depended on visibility of the vertebral border and of the angle of the scapula. The general height of the scapula was also noted; mobility may occur when muscles attached to the vertebral border of the scapula relax. In ectomorphic children, mobility is common as the muscles are relaxed and may be relatively underdeveloped.

(3) *Poking of neck* This was graded 0, 1, 2 and 3. Again, at nearly all ages a rating of 1 was so common that it could be regarded as normal. The children who were rated 'straight' neck on examination, (i.e. a neck which is roughly parallel to the line of gravity) were those with fairly well marked lordosis over-corrected by bracing back the scapulae and throwing back the head.

(4) *Poking of head* is uncommon; the head usually extends when the *neck* pokes forwards, so that eyes may be brought back to the horizontal position.

Round shoulders is a common finding during the early growth spurt; it is important to discover where the main fault lies and to make the child appreciate the importance of recognizing the fault, and of establishing a new body image (*see* Chapter 5).

Variations relating to the pelvic girdle and lower limb complex

Three conditions are relevant to this study; tight hamstrings have already been discussed in connection with the vertebral column; pelvic inclination and hyperextended knees are important factors in establishing balance and maintaining erect posture.

(1) *Pelvic inclination and lordosis*

The pelvic tilt was measured six-monthly using Wiles' inclinometer; data is scanty because these measurements were not introduced at the beginning of the study, but certain trends emerge (*Figure* 3.8).

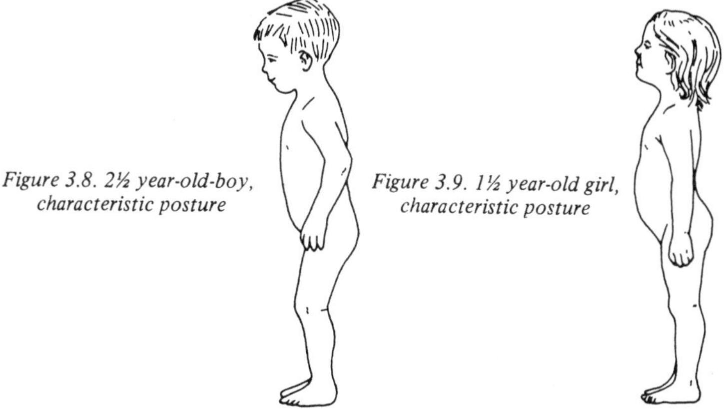

Figure 3.8. 2½ year-old-boy, characteristic posture

Figure 3.9. 1½ year-old girl, characteristic posture

In the second and third years the characteristic picture of potbelly and lordosis is seen; it is the child's method of distributing weight and ensuring balance. The pelvic tilt is very variable during these early years; it may be as low as 28 degrees or as high as 40. The child appears to vary the degree of lordosis by altering the curvature of the lumbar spine rather than altering the tilt of the pelvis, which is found on measurement to be lower than one would expect. He often maintains his balance by leaning forward and keeping his knees slightly bent (*Figure* 3.8).

The child who is seven years old (or over) (*Figure* 3.10) tends to tilt his pelvis and protrude his abdomen and also to hyperextend his knees, thereby distributing his weight evenly (antero-posteriorly) on both sides of the line of gravity. Thus during the middle school years we may expect to find a pelvic inclination between 30 and 40 degrees; in some children this figure varies (up or down) but in most the angle grows steadily smaller (*Figure* 3.11).

Fat children, especially those with a high degree of mesomorphy, with prominent abdomens, often manage their extra weight quite well without tilting the pelvis unduly; they do, however, tend to develop knock knee. Pads of fat may keep the condyles apart, or knock knee may be part of a balancing device comparable to that used by the toddler with developmental knock knee.

During the time that posture is stabilizing, the pelvic tilt is decreasing and is more consistent. When posture is fully stabilized (and this is associated with the growth spurt) the child's postural behaviour changes. During clinical assessment the child usually stands

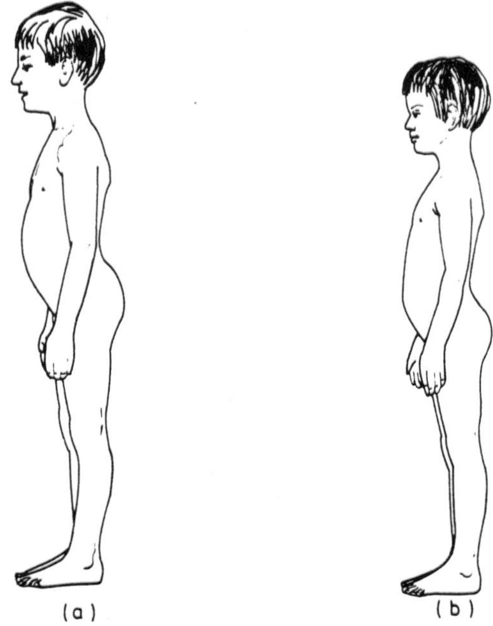

Figure 3.10. Lordosis and pelvic tilt in (a) seven-year-old boy; (b) seven-year-old girl

passively, almost stolidly, before the examiner; the pelvic tilt is remarkably constant, and is usually below thirty degrees; the knees are no longer hyperextended. By the time the boy or girl is approaching the age of 20, the pelvic tilt is fairly constant at 18 to 20 degrees.

Summary

Pelvic inclination is an important mechanism in maintaining balance in the growing child; it enables him to distribute his weight about the line of gravity when body proportions alter. It may vary from 25° to 40° in early years, is usually between 30° and 40° in mid-school period, and settles down to less than 30° during the growth spurt; it remains fairly constant at 20° from the age of 18 onward.

(2) *Hyperextended knees*

Hyperextended knees can often be observed in children between the ages of six and eleven; they may be regarded as part of a device used to adjust weight distribution antero-posteriorly. After the age

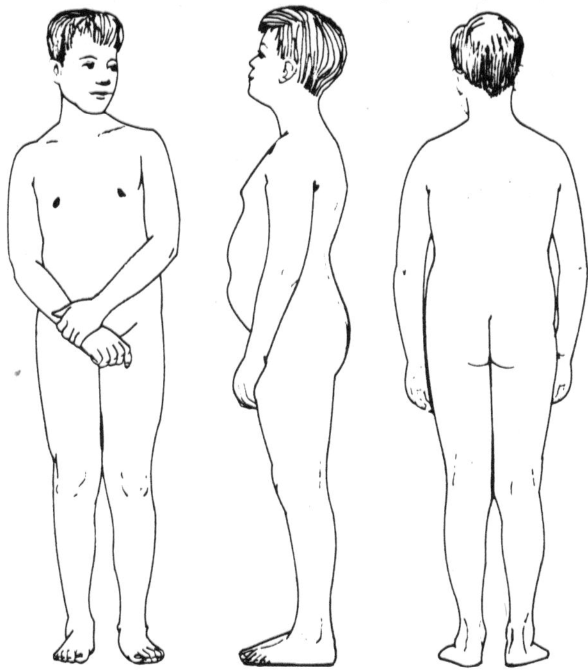

Figure 3.11. Boy aged thirteen years, pelvic tilt

of eleven, and before the age of six, children and adults tend to relax the knee joints slightly in position of rest.

REFERENCES

Asher, H. (1967). *Muscle and Bone,* London: Arlington Books

Hamerton, J. L. (1962). 'Bipedalism'. *Cytogenetics,* **15,** 99

Illingworth, R. S. (1972). *Development of Infant and Young Child,* London and Edinburgh: Churchill Livingstone.

James, J. I. P. (1951). 'Common spinal deformities in children'. *Br. med. J.,* **8.** 271

Krogman, W. M. (1951). 'The scars of human evolution'. *Scient. Am.,* (185(6), 54–57

Lambrinudi, C. (1934). 'Adolescent and senile kyphosis'. *Br. med. J.,* **2,** 800

Lead article (1964). 'Man's place in nature'. *Lancet,* **2.** 651

Le Gros Clark, W. (1962). *History of primates,* London: British Museum

Washburn, F. L. (1960a). 'Tools and human evolution'. *Scient. Am,* **203,** 63

–(1960b). 'The analysis of primate evolution with particular reference to the origin of man'. *Symp. quant. Biol.* **15,** 67

Wells, H. G., Huxley, J., Wells, C. P. (1946). *Science of Life,* London: Cassell

CHAPTER 4

4 Postural Patterns from Birth to Maturity

POSTURE OF THE CHILD *IN UTERO*

At first the fetus floats freely in his amniotic bath, changing his position when his mother changes hers. During the second half of prenatal life, however, space is limited, as he is unable to extend in any direction without encountering the uterine wall; he therefore adopts a position of maximum flexion, made possible by the presence of oestrogen (which has a relaxant effect) in the amniotic fluid. Each infant is said to acquire a 'position of comfort' in the uterus, which he prefers to all others, and which he continues to adopt (if allowed) during early postnatal life.

THE FIRST YEAR OF LIFE

The newborn baby, lying in his cot exhibits a large, gently kyphotic curve (*Figure* 3.3) extending from the cervical to the sacral vertebrae. If he is supported in a horizontal position, flexor tone appears to predominate in his limbs and muscle tone is more marked in the extensor muscles than in the flexor. If he lies in his cot in the prone position he soon tried to raise his head, and by the time he is about three months old will probably hold it erect. This is his first attempt to defy gravity. By the time he is sitting up he has a well marked compensatory spinal curve (convex anteriorly) in the cervical region, which results from modification of the intervertebral discs.

Should the child spend his first months in the prone or in the supine position? Opinions differ (*see* lead article *Br. med. J.* 1961) The average age at which infants begin to roll over completely from the prone to the supine position is about five months, and from supine to prone at about six months; thus, the child is unable to choose his own position before he is six months old. In England, most children are placed in their cots in the supine position; in many parts of the United States it is customary for them to lie prone, without pillow, from the outset. Children reared in this way learn to extend their heads early, and therefore will have a less restricted view of the world around them than those who lie on their backs gazing at the ceiling or the sky.

Holt (1960) showed that Iowa babies accustomed to sleeping prone showed more advanced motor development in the prone position than English babies who had been reared in the supine position. The 'average' child sits unsupported for the first time between the ages of six and eight months. During the period in which he is learning to sit and stand, a compensatory curve appears in the lumbar region; there are now two primary curves (concave anteriorly) and two secondary curves (convex anteriorly) (*Figure* 3.4). The child may make serious attempts to walk by the end of the first year (*Figure* 4.1).

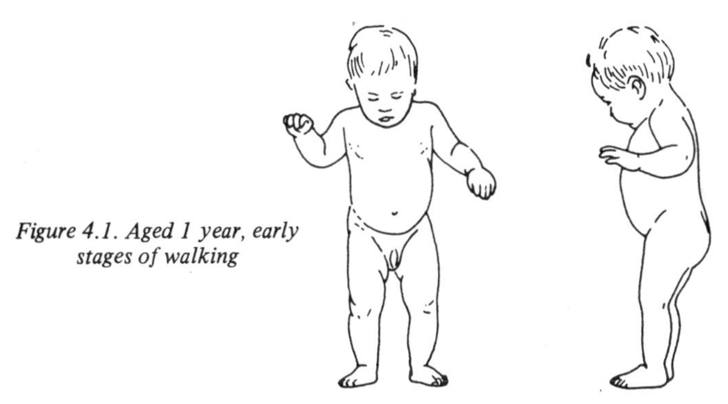

Figure 4.1. Aged 1 year, early stages of walking

THE SECOND YEAR: STANDING AND WALKING

During the second year the child is slowly acquiring the ability to stand up straight, and to balance, both laterally and antero-posteriorly. He stands and walks on a broad base, with legs wide apart; the distance between the upper thighs is made greater by napkins. His pelvic tilt varies; his abdomen bulges. He may show some lordosis; he leans forwards, legs partly flexed at knees, arms abducted and slightly flexed at elbows, as though he is steadying himself with partly unfolded wings. He totters cautiously from chair to chair until, when postural reflexes are established, he masters the arts of propulsion and of balancing while standing more or less erect. When he reaches the age of two, we can expect him to walk and run; he may appear to be flatfooted, his legs will be nearer together, but lordosis and prominent abdomen may still be evident (*Figure* 4.2).

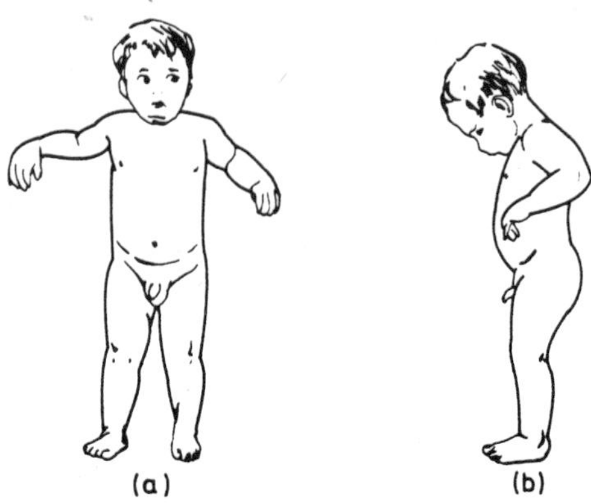

Figure 4.2. Age 1½ years; (a) walking on a broad base (b) learning to balance

AGE 2 TO 6 YEARS

This is the knock knee period (*see* Chapter VI). During this time the

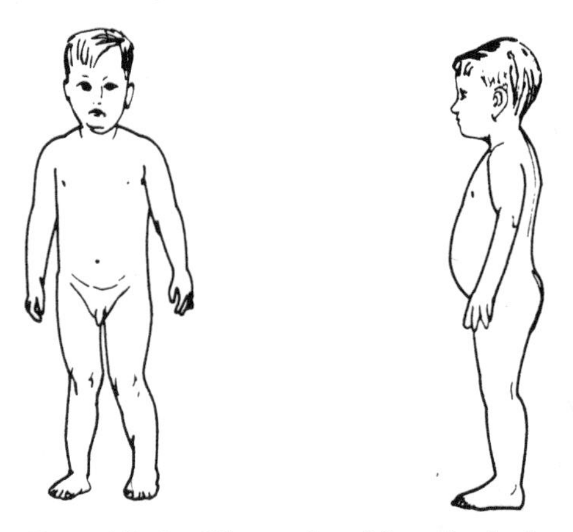

Figure 4.3. Age 2½ years, knock knee developing

knees approach each other, but a broad base still appears to be necessary for lateral balance and is maintained by torsion of the tibiae. The child is becoming accustomed to the upright posture;

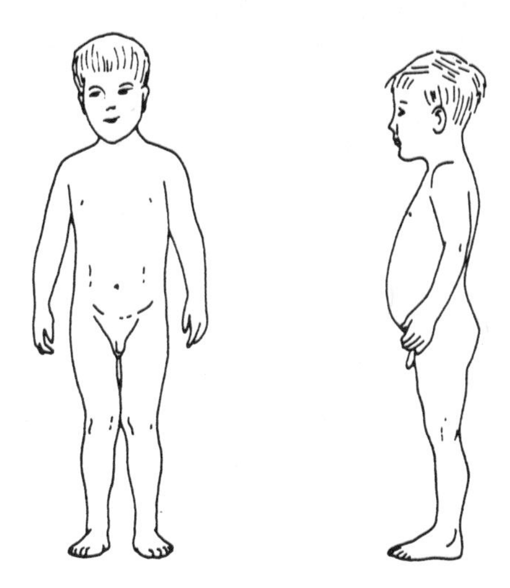

Figure 4.4. Age 3½ years, knock knee still present

the abdomen is not so prominent and, towards the end of the period, the foot usually shows a well developed longitudinal arch (*Figures* 4.3, 4.4).

AGE 6 TO 12 (BOYS), 6 TO 10 (GIRLS)

This includes the greater part of the primary school period. At six most children have corrected their knock knees (if they had them). Their height is increasing steadily though velocity remains constant. Before puberty the limbs grow more rapidly than the trunk but. with the general increase in velocity, the trunk begins to grow faster. At puberty the rates of growth of trunk and extremities are about equal, but the trunk continues to grow after the extremities have slowed their rate of growth in the postpubescent period. These differences in segmental growth rate cause changes in the ratio of sitting height to total height. The sitting height represents about 70 per cent of total length at birth, but falls rapidly to about 57 per cent at three years, and to about 52 per cent at 14 years (boys) and 16 years (girls). Thus proportions have been altered, and new adjustments must be made to meet the demands of gravity.

The most obvious characteristic of the child's posture at the beginning of the primary school period is his extreme mobility. If asked to stand upright, he may rotate his head, alter his pelvic tilt and, possibly, hyperextend his knees. His arms may assume various positions, and he often appears to be 'doing' foot exercises, curling his toes under and throwing most of his weight on the outer side of his foot. If asked to exercise (running or jumping) he may respond readily, tilting his trunk forwards, and his pelvis forwards and downwards; consequently his buttocks may protrude, and he may look very much like an ape, particularly if he happens to be high in ectomorphy, with long skinny legs. When the exercise is over, he takes some time to settle down; he appears to be making attempts to adjust to gravity, and finally he comes to a solution, though he has not yet a definite body image (*see* Chapter 5) to which he can return. This description applies to children at the lower end of the age group; as years pass posture tends to become more stable.

STABILIZATION OF POSTURAL PATTERN

Changes in proportion which occur during the adolescent spurt necessitate further adjustments to gravity. The pelvic tilt may have decreased to 25 or 30 degrees; the knees are often slightly bent; hyperextension of knees is no longer necessary in order to balance prominent abdomen. The lower segment has stabilized itself in readiness to carry the adult frame; mobility is still apparent in the upper segment.

Thus, in the primary school child, we have a period of intense mobility at the lower end of the age range (from six onwards) and a more static period at the upper end (10 years in girls, 11½ years in boys). It will be seen that this static period appears to coincide with the beginning of the adolescent general spurt in growth. Stabilization of postural patterns is slowly taking place; and various attempts to come to terms with gravity can be observed.

In order to study these postural patterns the following observations were made during clinical examination (*Figure* 2.4). Position of head, neck and shoulders, scapula, prominence of abdomen, lordosis, kyphosis, pelvic tilt and postures of knees were noted, before and after exercise. Two typical postures will be described.

Posture 1: Girls, primary school period

(a) Age 7½; pelvic tilt 40 degrees
(b) Age 8; pelvic tilt 35 degrees
(c) Age 7; pelvic tilt 42 degrees
(d) Age 12½; pelvic tilt 36 degrees

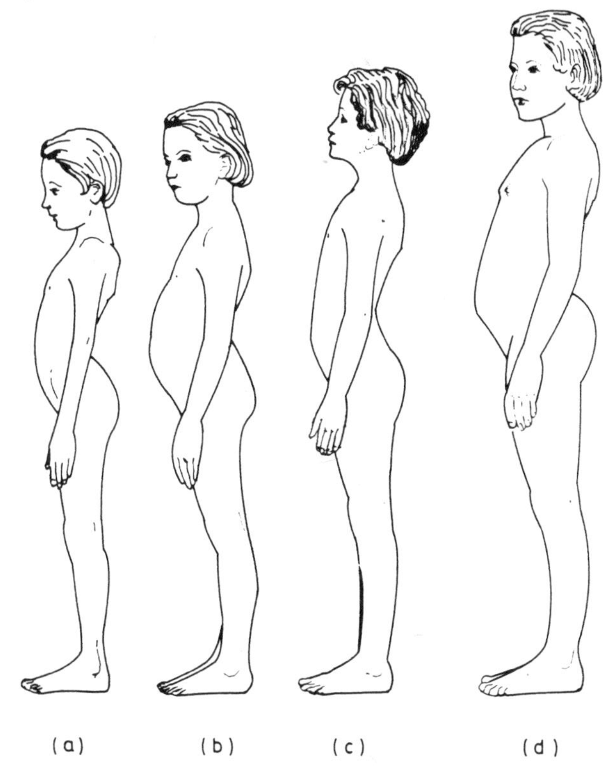

Figure 4.5. Posture 1: girls, primary school period

This posture (*Figure* 4.5) is often observed in the enthusiastic mesomorph. It is a positive posture; the child stands with pelvis tilted downwards and forwards and this may be associated with lordosis and mild prominence of the abdomen. To maintain balance, the knees are hyperextended, the scapulae may be braced back by the trapezius, and the spine slopes backwards, either from the lumbosacral joint or from the upper cervical spine. The spine and scapulae may be tilted back as far as a vertical line tangential to the curve of the buttocks, or even further. Lordosis may be extreme and a grotesque posture may be produced; the head is held up and the neck may be straight, i.e. vertical in direction.

Posture 2: Girls, primary school period

(a) Age 6; pelvic tilt 58 degrees
(b) Age 8; pelvic tilt 40 degrees
(c) Age 10½; pelvic tilt 38 degrees

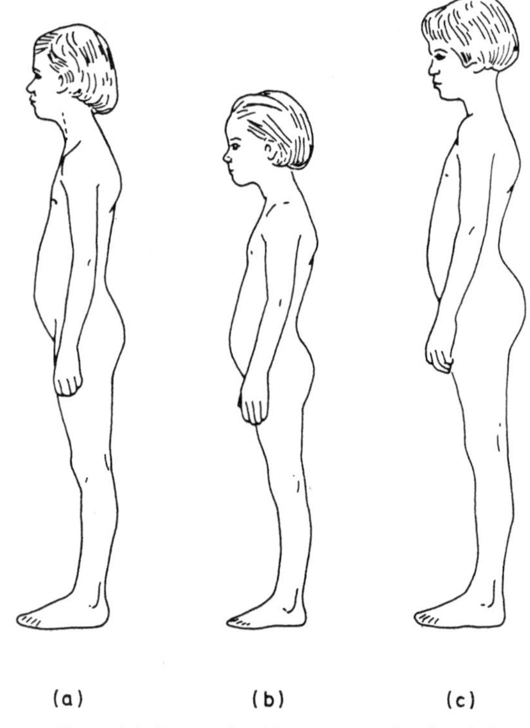

Figure 4.6. Posture 2: girls, primary school period

This postural pattern (*Figure* 4.6) is common in children high in ectomorphy. It is more passive than Posture 1 (*Figure* 4.5), as if the child were making very little effort to combat gravity. The pelvis is tilted, and the abdomen may protrude slightly; the knees may be hyperextended. Lordosis may be slight; the whole spine may lean back rather sharply from the lumbosacral joint. The scapulae are winged, the shoulders are held in the forward position; the neck is directed forwards and there may be a mild postural kyphosis. This pattern resembles closely the one described by Wiles and others (1946) as 'swayback'; they regarded it as a 'defect'. The 'swayback' pattern was very common in the group of children investigated at the primary school level, but was rarely seen in the adolescent.

Posture 1: Boys, primary school period

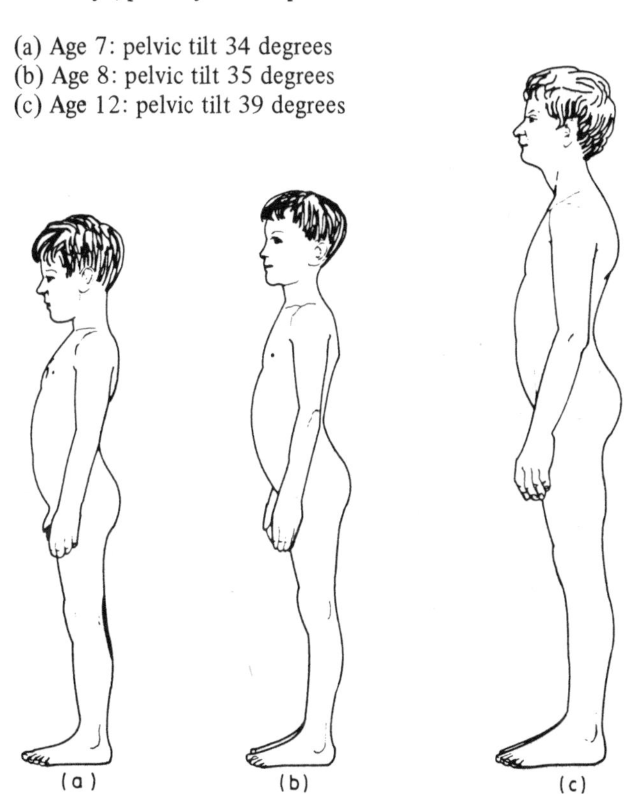

Figure 4.7. Posture 1: boys, primary school period

These boys (*Figure* 4.7) show the characteristics of Posture 1 (boys) described in text, i.e. prominent abdomen, lordosis, and hyperextended knees; Their pelvic tilts measured by inclinometer, lie between 34 and 39 degrees.

Posture 2: Boys, primary school period

(a) Age 8½; pelvic tilt 32 degrees
(b) Age 9; pelvic tilt 30 degrees

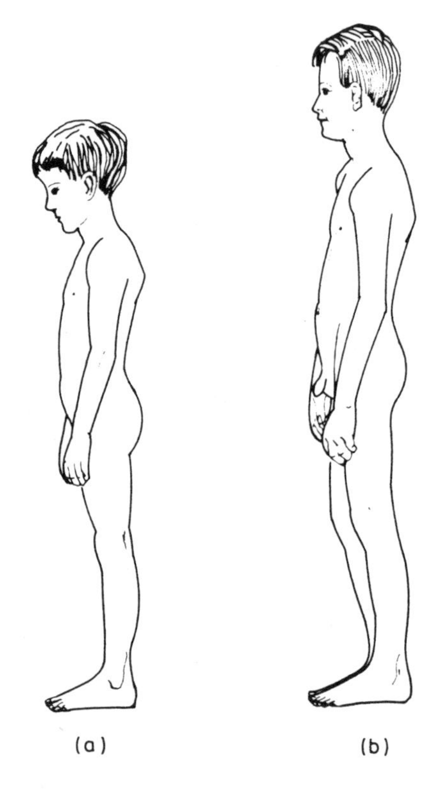

Figure 4.8. Posture 2: boys, primary school period

These boys are learning to balance without increasing pelvic tilt or hyperextending knees.

Posture stabilizing; abdomen no longer protruding, lordosis absent.

(a) Girl aged 20; pelvic tilt 30 degrees
(b) Boy aged 12

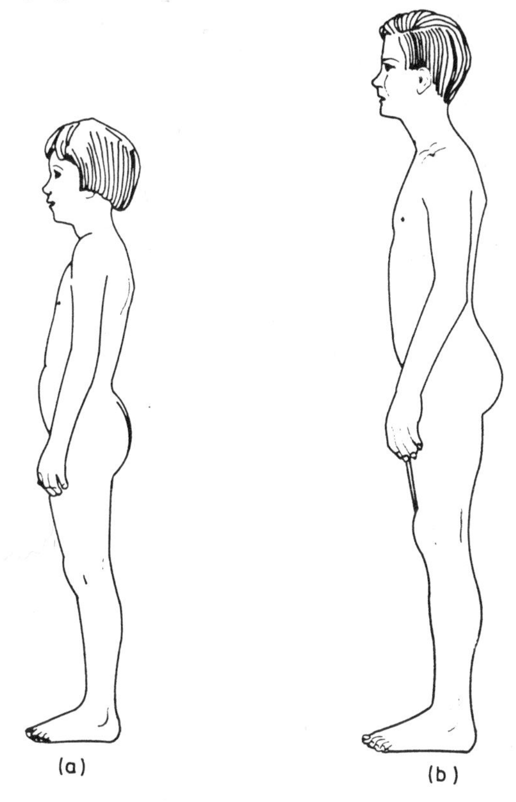

Figure 4.9. Posture stabilizing

'Ideal' posture

(a) Girl aged 10½
(b) Girl aged 13½

In Chapter 5 the question of ideal posture is discussed. Posture was considered (by some observers) to be ideal if the line of gravity passed through the mastoid, the acromion, the socket and a point just anterior to the ankle joint. The girl depicted here approached the ideal

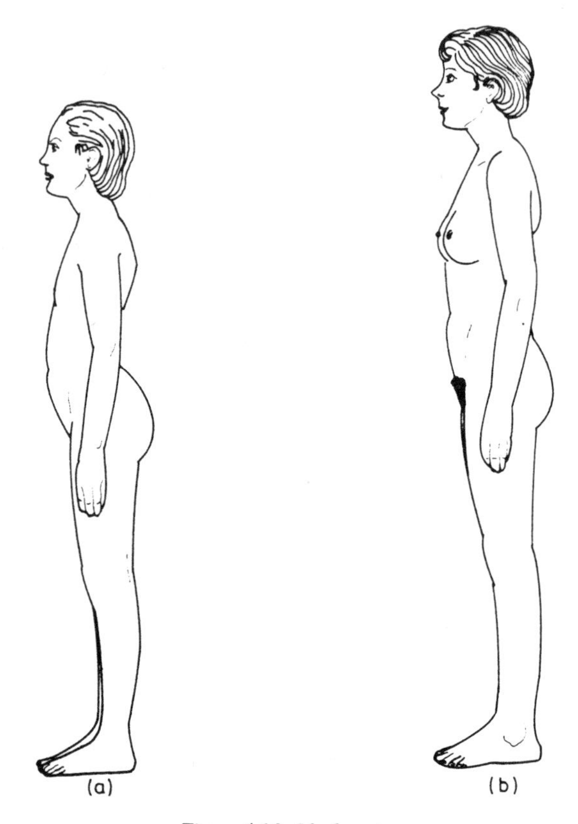

Figure 4.10. Ideal posture

more nearly than any of our subjects. At 10½ her abdomen was prominent and her pelvic tilt was 28 degrees; at 13½ her abdomen is flat, her pelvic tilt is 38 degrees and she has a static, peaceful pose.

Summary

Growth is steady during the primary school period; the pelvis is tilted (30–40 degrees) and is balanced by hyperextended knees and bracing back of the vertebrae as in Posture 1. In posture 2 pelvic inclination is not so marked as in Posture 1, and the child may balance by extending the lumbar spine; knees show less hyperextension.

ADOLESCENCE

The child no longer wriggles, and, after exercise drops naturally into his habitual pose, based on his body image. There is very little change in his lower segment after exercise; the pelvic tilt is nearly always near 20 degrees and the buttocks no longer protrude. His actual stance will depend on body build and stage of development. He now emerges as an individual; it is not possible to describe a typical adolescent posture.

DIAGNOSIS AND SIGNIFICANCE OF POSTURAL VARIATIONS

In this chapter, in which developmental changes from birth to maturity have been described, no suggestions for remedial or orthopaedic treatment have been made. Many conditions formerly regarded as defects are now considered to be developmental in origin, and to be within normal limits. For example, if 'flat feet', bow legs, knock knee, round shoulders and lordosis occur outside the normal age range, or if they are exaggerated, orthopaedic advice should be obtained.

Treatment at adolescence may be necessary if a faulty posture has been assumed which is based on a faulty body image. The adolescents in this study usually showed a static lower segment, but not infrequently showed the 'round shoulders' syndrome with head extended, neck forwards and shoulders forwards. Barlow (1955, 1956) is of the opinion that the posture of the head is the most important single factor in extablishing good posture, and, if the position of the head is correct, the rest of the body will fall into line. In any case the only form of treatment which might improve these children would be recognition of body image and re-education.

REFERENCES

Barlow, W. (1955). 'The psychosomatic problems in postural re-education'. *Lancet,* **132,** 659

–(1956). 'Discussion on postural re-education'. *Proc. R. Soc. med.,* **49,** 667

Holt, K. S. (1960). 'Diaper rash, self inflicted excoriation and crying in full term newborn infants, kept in prone or supine position'. *J. Paediat,* **1,** 57, 884

Leading article (1961). 'Prone or supine'. *Br. med. J.,* **1,** 1304

Wiles, P. (1949). *Essentials of Orthopaedics,* London: J. & A. Churchill

CHAPTER 5

5 Physiology of Posture

Posture may be defined as the position of the body in space, with special reference to its parts. Carriage, attitude, and pose are sometimes used synonymously with posture; a pose, however, is assumed voluntarily for photograph, portrait or exhibitionism; and an attitude may express an emotion, such as fear and aggression. Carriage usually refers to gait. Poses, attitudes and carriage are transitory events. Posture or the 'postural set' in the adult is a permanent habit of standing, to which he returns after exercise; it is a position of rest; it is characteristic of the individual, and it is probable that it depends on the body image.

It is difficult to describe what is meant by 'body image' since those who use the term (in speech or writing) do not always agree amongst themselves as to its exact meaning. Several groups of workers have used the term body image and it might be worthwhile to discuss briefly how far they agree in their interpretation of its meaning.

The neurologists

Critchley, writing in 1950, said that only in the last two decades had neurological phenomena been related to body image and disorders. The expression body image, he said, refers to a 'mental idea which a person possesses regarding his own body and its physical and aesthetic attributes'.

Head and Holmes (1911; quoted by Critchley, 1950) were the first English observers to draw attention to the idea of body image but many French neurologists had, before 1911, discussed the body image and kindred subjects such as phantom limbs. Many factors contribute to formation of body image, the chief ones being visual, tactile and proprioceptive. Head and Holmes preferred to call the body image the postural model; they pointed out the indestructibility of the body image, e.g. it is resistent to mutilation and persists in phantom limbs after an amputation.

Lhermitte (1939; quoted by Critchley, 1950) considered that the body image was built up in childhood; important factors in its production were pain, stimulation and play. The body image is very plastic in early days; this is evident from children's drawing with their anatomic distortions, preoccupations with genital organs and with pornographic graffiti. It is some years before body image stabilizes and conforms to adult pattern.

The body image in infancy

For the first weeks of life the baby is just passing time; he soon however starts in the serious business of perception and is busy constructing his own space capsule. Through this capsule he gets his limited self knowledge through EYE and HAND inspection. First he notices space on either side of body and then he investigates space and objects in midline.

Later, he turns to middle line so that he can inspect his hands preparatory to picking up objects with both hands, touching hand to hand.

In addition to early progress in perceptual studies, the infant must become aware of his body, and the whereabouts of his body and limbs.

'Tasks' to heighten body awareness (older children)

Trampolining. This is an activity through which the objective body image may be strengthened. As the body becomes weightless, the child rising above the trampoline, it is suggested that heightened awareness is obtained relative to two axes around which the body may move. He begins to localize his body parts; postural awareness occupies a key position in the idea of oneself.

The body image; physiotherapists

The idea of the body image, or, as some prefer to call it, the postural model or schema, is important in re-education of children with postural disorders. In the past, there has been much emphasis on effort; but words or ideas are only effective if, at the same time, demonstrations of sensory experience to which they refer are given (Barlow, 1955).

There must be a period of 'conditioning' in which the new model (verbal or otherwise) is trained. The new model is associated with muscle tension not only at rest but in preparation for movement (Barlow, 1956). In this way the child achieves a basic resting state of postural equilibrium which he can apply at will.

When the child assumes the erect position the stretch reflex (Barlow, 1956; Magoun, 1963) comes into play as the foot touches the ground. As the muscle is stretched, muscle spindles lying between longitudinal muscle fibres are also lengthened and send information to the cord by the afferent nerves. In the cord connection is made with the large alpha motoneurone, which causes the appropriate muscle to contract. At the same time, by means of the Renshaw cells small collaterals connect with afferents which serve the muscles

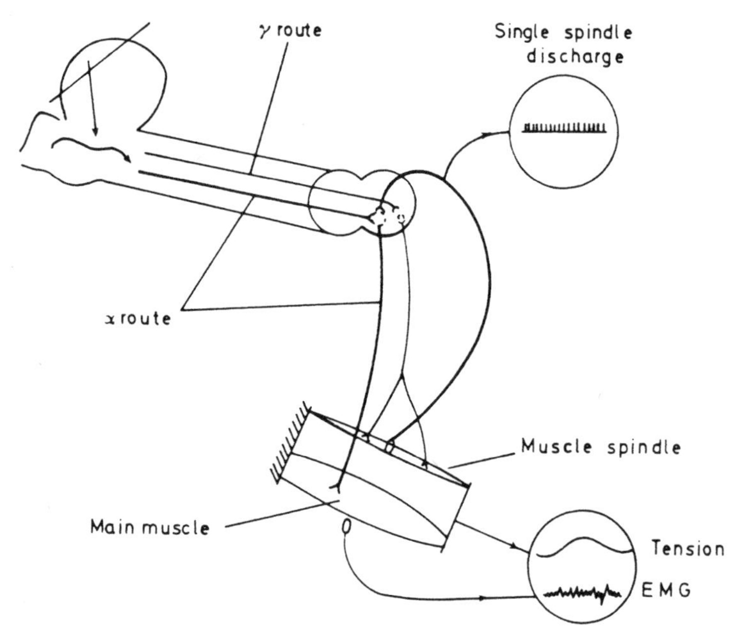

Figure 5.1. Diagram of gamma feedback loop between the spinal cord and muscle, together with the cerebellar control of alpha and gamma motor discharge From Granit, Holmgren and Merton (1955)

concerned in the supporting reactions, i.e. the antagonists and synergists. When the muscle, excited by the efferent neurons, contracts, the spindle is also shortened by virtue of its anatomical position between the muscle fibres.

The muscle spindle consists of two contractile poles, separated by sensory nerve endings. When it shortens, the sensory body sends information through the afferent gamma route to the cord, and thence up to the cerebellum. From the cerebellum impulses travel by the gamma route to the cord and connect with the gamma motoneurones, which supply the contractile poles in the muscle spindle. By this servo-loop mechanism the length of the muscle fibres can be controlled, and constantly modified as need arises, and posture and tone can be maintained. Other afferent stimuli concerned in maintaining posture are neck reflexes, visual and auditory stimuli and the vestibular apparatus.

What keeps the body erect? It is not easy to stand stock-still. Swaying occurs, and this must be constantly corrected by appropriate muscle contraction; the feet must be planted firmly on the ground, and the body must be balanced on the ankle joint. As the line of gravity lies in front of this joint there is a tendency to fall forwards which is counteracted by contraction of the calf and lower back muscles. It is possible that the capsules and ligaments of the joints and the deep fascia are in a state of passive tension and that they may assist in the maintenance of erect posture.

Electromyographic studies (Joseph, 1960) have shown that, when one stands erect, muscular contraction occurs only in the calf muscles and lower back muscles, i.e. there is just enough tension in the soleus to prevent the body from falling forward; the position of the rest of the body is held by ligamentous tension and balance.

The above discussion refers to adults. While postural patterns certainly occur in children, they change continually according to the degree of development at which the children have arrived. It is not until they arrive at the age of ten or older that a constant pattern emerges.

Gravity

If the upright position is to be maintained, the centre of gravity of the whole body must be above the area occupied by the feet. Brain (1959) points out that without conscious thought we are able to balance our not inconsiderable bulks on two surfaces of a few square inches each.

The centre of gravity of a body may be defined as a fixed point in the body through which the resultant of the gravity forces acting on all the molecules of the body may be said to act. It may be determined by suspending the body in question and allowing it to hang freely; the centre of gravity must lie somewhere on a vertical line passing through the point of suspension. By choosing a second point of suspension, another vertical line can be obtained; the centre of gravity must lie at the point of intersection of these two lines. These vertical lines are known as *lines of gravity*. Obviously this method of determination of the centre of gravity cannot be applied to living and moving creatures; but it is generally agreed that when 'good' or ideal posture obtains, the line of gravity passes through the following points; mastoid, tip of shoulder, hip, front of ankle (Burt, 1950).

Good, bad, and ideal postures

For the purpose of this discussion, it would be as well to regard Burt's definition (1950) as an ideal posture, rarely achieved. Appleton (1946) has suggested that good posture is one that functions well, that resists disability and looks well. Others have suggested that it is one in which the minimum of muscle activity is used to maintain the body in a state of equilibrium. For example, Tucker (1960) describes an active alerted posture, which is the result of an attitude of mind towards body, promoting mental and physical equilibrium and poise. Barlow (1955) points out that the old idea of faulty posture is associated with maldistributed tension because over-contraction of muscles diminishes spindle activity; consequently there is absence of feedback, i.e. information about degree of postural deformity is not fed back into the brain. Postural errors must be made conspicuous to the patient so that he can recognize them and return 'at will' to balanced resting state.

It is obvious that it would be impossible to mould children into a stereotyped posture. Postural patterns vary frequently in children under ten years old; they are constantly experimenting with different ways of reacting to gravity. Adolescents and adults, on the other hand, usually have acquired well marked postural sets which may or may not be regarded as satisfactory.

Our general aim for adolescents should be to promote a posture which is most economic of effort, and will therefore prevent undue fatigue. We want to encourage patients to balance their bodies over the foot area, so that anti-gravity muscles are not over worked. Tucker (1960) points out that upright posture can be maintained with very little muscle activity except in two areas, i.e. in the calf muscles which stabilize the ankle joint, and in very slight contraction in erector spinae. These muscles contract slightly to control swaying. The weight of the body is counterbalanced by passive tension in joint tissue, i.e. in fasciae and joint capsules.

REFERENCES

Appleton, A. B. (1946). 'Posture'. *Practitioner,* **156,** 48
Barlow, W. (1955). 'The psychological problems of postural re-education'. *Lancet,* **132,** 659
–(1956). 'Discussion on postural re-education'. *Proc. R. Soc. med.,* **49,** 760
Brain, R. (1959). 'Posture'. *Br. med. J.,* **1,** 1489
Burt, H. A. (1950). 'Effects of faulty posture'. *Proc. R. Soc. Med.,* **43,** 187
Critchley, M. (1950). 'The body image in neurology'. *Lancet,* **1,** 335
Granit, R., Holmgren, B. and Merton, P. A. (1955). 'The two routes for excitation of muscle and their subservience to the cerebellum'. *J. Physiol.* **130,** 213
Head and Holmes. (1911). Quoted by Critchley, *see* above
Joseph, J. (1960). *Man's Posture: Electro-myographic Studies.* Oxford: Blackwell
Lhermitt. (1939). Quoted by Critchley *see* above
Magoun, H. W. (1963). *The Waking Brain.* Springfield, Illinois: Thomas
Tucker, W. E. (1960). *Active Alerted Posture.* London: Livingstone

Additional reading:

Cratty, B. J. (1967). *Movement, Behaviour and Motor Learning.* Philadelphia: Lea & Febiger
Walsh, E. G. (1957). *Physiology of the Nervous System.* London: Longmans

CHAPTER 6

6 Knock Knee (Genu Valgum)

DEFINITION

Knock knee is defined as a condition in which the medial malleoli do not meet when the child sits on couch or floor with legs extended, medial femoral condyles touching and patellae facing upwards. The degree of knock knee present is estimated by measuring the distance between the medial malleoli (*see also* page 61 on the tibiofemoral angle).

INCIDENCE OF KNOCK KNEE

In the study described in this chapter the degree of knock knee present in 185 children was graded six-monthly as follows:

Less than 2·54 cm: 0
2·54 cm to 5·08 cm: 1
More than 5·08 cm: 2

Figures 6.2 and 6.3 show that knock knee is common at the age of 3½ years; its incidence decreases rapidly between the ages of five and six; very few cases are seen in eight-year-olds.

Morley (1957, *Figure* 6.4) made 1000 examinations of normal children from infant welfare centres in Northwest London; she found knock knee to be commonest in children between the ages of 3 and 3½ years. Twenty-two per cent of children in this age group had a malleolar separation of two inches or more, and only 1·2 per cent of children aged seven years and over had an equivalent degree of knock knee. After the age of 3½ the incidence declined. Thus the same general pattern emerged in both studies. It is now generally recognized

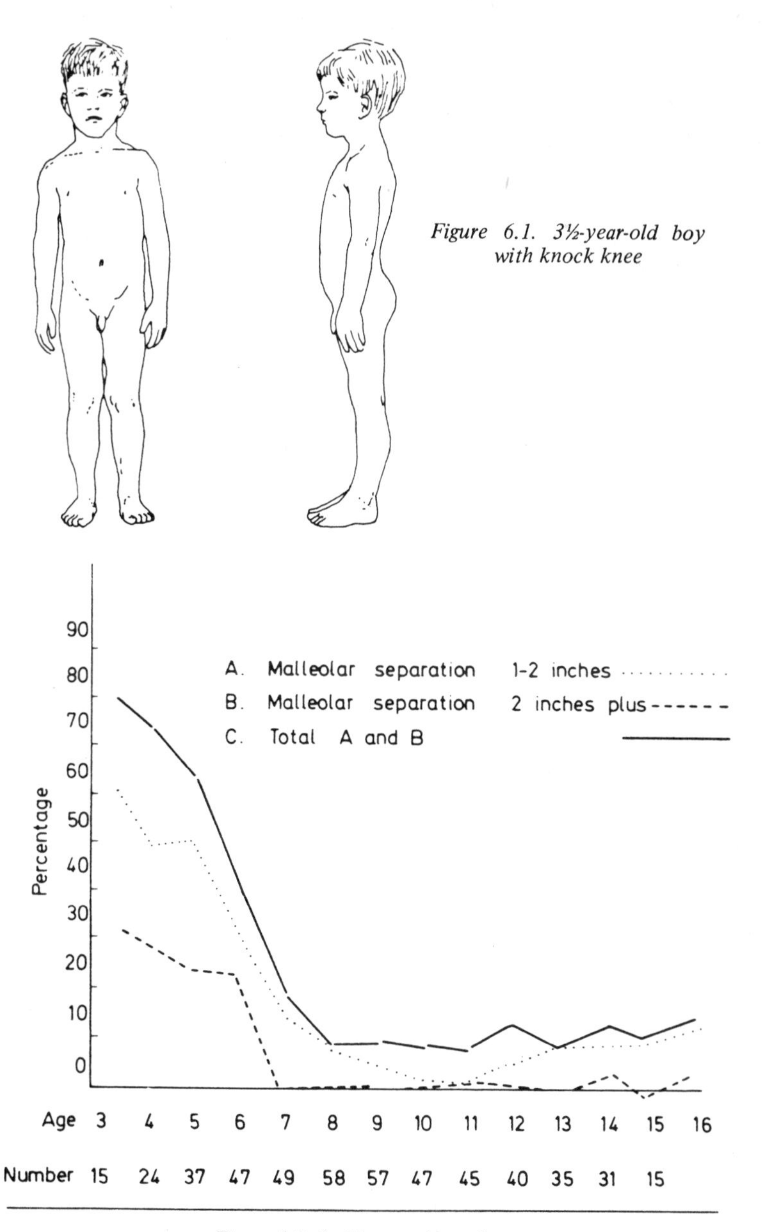

Figure 6.1. 3½-year-old boy with knock knee

Figure 6.2. Incidence of knock knee; boys

that most cases of knock knee are developmental in origin. Morley regarded knock knee as a developmental deviation, usually needing no treatment; I should go further and regard it, not as a deviation, but as a normal phase of development.

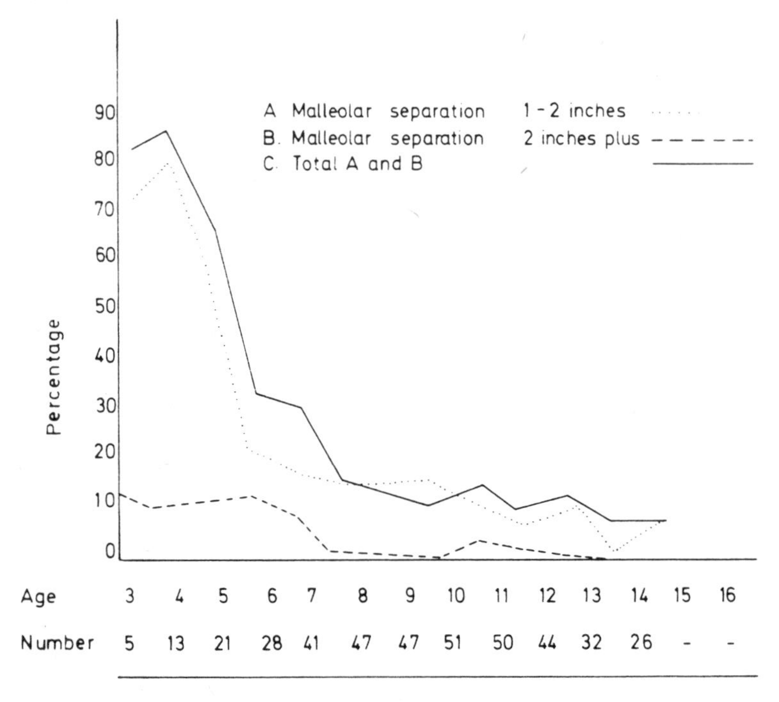

Figure 6.3. Incidence of knock knee; girls

Figures 6.2 and 6.3

In these figures the percentage of children with knock knee is plotted against age; the figures were obtained from serial clinical examinations carried out for longitudinal studies.

The numbers of children studied before the age of four were too small to give significant results. Three curves were plotted for each sex:

(A) Malleoli separated by a distance of 2·54–5·18 cm.
(B) Malleoli separated by a distance greater than 5·18 cm.
(C) Malleoli separated by a distance greater than 2·54 cm (A & B).

In both sexes the incidence of knock knee is seen to fall rapidly between the ages of four and eight. Girls continue to show a mild

Figure 6.4. Incidence of knock knee (Morley, 1957. Reproduced by permission of the Editor, British Medical Journal.)

degree of knock knee longer than boys, and a small rise in incidence is apparent in girls between the ages of ten and fourteen.

NATURE OF KNOCK KNEE

During the early part of the century, knock knee was generally regarded as a rachitic manifestation. When the incidence of rickets decreased in this country during the Second World War, as a result of improved food distribution and better appreciation of food values, the incidence of bow legs declined dramatically; but knock knees were still frequently seen, though most of them were mild in degree. It became obvious that rickets could no longer bear the blame for all cases of knock knee.

Unequal growth of the femoral condyles was suggested as a possible cause for knock knee. It was supposed that the medial condyles grew downwards, thus altering the tibiofemoral angle, and causing separation of the malleoli; and that later the lateral condyles grew downwards, thus restoring the original direction of the tibiae, enabling the malleoli to meet and correcting the knock knee. Knock knee was regarded as a deformity, at one time, and often received treatment. The idea that it was developmental in origin and usually self-correcting gradually emerged. It was still thought by many, however, that the mechanism of production of knock knee in early stages depended on downward growth of medial condyles.

METHOD AND PURPOSE OF STUDY OF DEVELOPMENTAL KNOCK KNEE

This study aims at discovering how and when developmental knock knee arises, and how and when it is corrected. The radiological changes which can be seen in the lower femoral epiphysis during the knock knee period are described and shown diagrammatically.

Six-monthly examinations were carried out. Children were examined in the long-sitting position, on the floor, with knees extended and pressed to the ground, with medial condyles touching, and patellae facing directly upwards. The distance between the malleoli was measured in centimetres.

ANATOMY AND DEVELOPMENT OF THE KNEE JOINT

The knee joint is well fitted for the functions of locomotion and weight-bearing; it is primarily a hinge joint formed by the articulation of the lower end of the femur with the upper end of the tibia. The fibula articulates with the lateral aspect of the head of the tibia, and the patella lies against the lower end of the femur. The joint permits flexion on sitting, and extension on standing; slight rotary movements can take place on full extension; the inner side of the femur continues to move after the outer side has stopped, producing a screw-like action which locks the joints and turns the lower limb into a stable weight-bearing column. The joint is unlocked by the popliteus, which rotates the femur in the opposite direction.

In the adult, the medial femoral condyle is more prominent than the lateral, because of the obliquity of the shaft of the femur; the lateral condyle is more directly concerned with weight-bearing and is therefore stronger than the medial condyle. The femoral shaft makes an angle of about 9 degrees with the vertical.

The ligaments

The internal lateral ligament is a flat, broad, triangular fibrous sheet, extending from the medial femoral condyle to the tibia. The external lateral ligament is a strong, rounded, deep, fibrous cord, extending from the lateral epicondyle to the head of the fibula.

The femur: torsion

Torsion of the femur occurs at intervals in both fetal and postnatal life (Frazer, 1965).

In the third week of fetal life the limbs begin to make their appearance as small buds in lateral ridges on either side of the trunk. The limbs are at first directed nearly backwards, parallel to the long axis of the trunk; by the sixth week their long axes are at right-angles to that of the trunk, and the three chief divisions of the limbs are marked off by furrows. The limbs next undergo a rotation of 90 degrees round their long axes, rotation being effected almost entirely at the limb girdle. The limb is rotated inwards and backwards, so that the tibial (pre-axial) border of the limb is directed medially and the flexor surface is turned backwards.

In postnatal life internal torsion shows itself clinically as a forward twist of the lower end of the femur, bringing the lateral condyle further forward, and pushing the medial condyle inwards and backwards.

Roberts (1962) defines the *angle* of torsion as the intersection of lines drawn through main transverse axes of upper and lower ends of the femur (*see Figure* 6.5). At birth this angle lies between 30 and 60

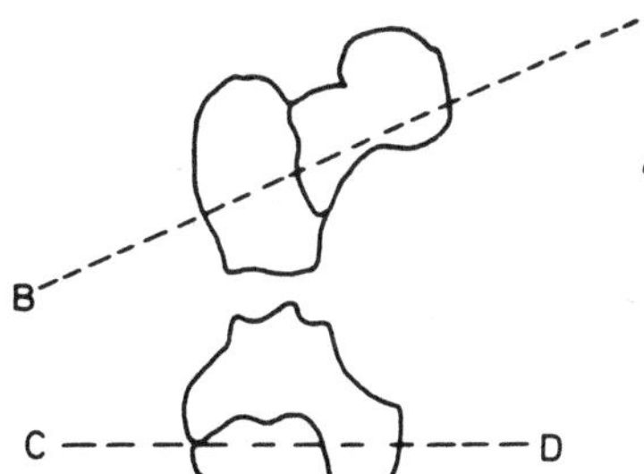

Figure 6.5. Right femur; anterior view. The lower end of the femur is twisted backward, i.e., it rotates in a backward direction so that vertical plane that contains CD is at an angle with vertical plane containing AB

degrees; it decreases steadily in postnatal life; at puberty it is 25–30 degrees and in adult life 12 - 15 degrees.

Appleton (1964) points out that, though the variations in form are such that the femur may appear to be twisted on its long axis, the mechanism is probably to be found in altered lines of growth in epiphysial lines, and the term torsion should not be understood as a literal twisting but rather as a description of shape.

When the child first assumes the erect posture the femur acquires a gentle bend with concavity backwards; when he begins to walk internal torsion decreases quite rapidly. Children who walk early are more likely to turn their toes in than those who walk later; this may be accounted for by a higher degree of internal torsion in the younger child.

As torsion of the femur decreases, the medial condyle moves forward and medially and the lateral condyle moves backwards; increase in length of the femur is also taking place. The site of the intercondylar fossa is apparent as a well marked depression in the popliteal space, and its change in position with growth and decrease in torsion can easily be traced.

Roberts (1962) points out that torsion of the femur plays an important part in postural changes; it may be associated with medial rotation of fetal limb (*see* above), or with other postural changes before birth, or it may be due to activity of the iliopsoas. Torsion after birth is closely connected with growth and weight transmission, and with postural changes, either as cause or effect.

The tibia

The tibia transmits the weight of the body to the ground; the upper end is massive and broad, and articulates with the femoral condyles. The shaft shows a sharp curved crest anteriorly, immediately beneath the skin. Unlike the femur, the tibia can undergo both internal and external torsion and therefore appears to twist on its long axis both inwardly and outwardly. Very little tibial torsion is seen at birth; the average amount is 4 degrees. During the postnatal period, the average degree of torsion is 20 degrees, so that a value of 0 degrees indicates internal torsion, and a value of 40 degrees denotes external torsion.

During the second year of life, when the child is learning to walk, in-toeing may occur. This may be due to internal torsion of the tibia, but is more often due to internal torsion of the femur (*see* above). Other causes of in-toeing include adduction of forefoot (mild) and metatarsus varus.

During the third year, external torsion of the tibia may occur; this has the effect of producing a wider walking base, and helps the top-heavy two-year-old in lateral balancing. The distance between the internal malleoli may be very slight; if the separation amounts to more than 2·54 cm when the child is sitting on the floor with legs outstretched, developmental knock knee is said to be present.

Between the ages of five and seven, when a wide walking base is no longer necessary, internal torsion of the tibiae may occur; and the malleoli may meet again, thus correcting knock knee (if it were present).

All children do not show the clinical picture of genu valgum, but, in most cases, a transient separation of malleoli (perhaps 2·54 cm or less) can be seen at some time during the third year.

The tibiofemoral angle

Measurement of the distance between the malleoli (*see* page 54) is not the only criterion for the diagnosis of knock knee; the degree of angulation between the femur and the tibia is also of importance, particularly in more severe degrees of knock knee and in those cases which fail to correct between the ages of five and seven. The lines which form this angle are drawn to the tibial tubercle as apex, one along the femoral shaft, and the other along the tibial crest. Most children with mild developmental knock knee have a tibiofemoral angle of 10 degrees (or 170 degrees). If the angle increases it is possible that spontaneous correction may not occur, and orthopaedic treatment may be necessary.

It is obvious however that, if knock knee is present, as the tibiae grow in length separation of the internal malleoli will increase. If the amount of separation does not increase, or is decreasing, correction must be taking place. Serial inter-malleolar measurements, therefore, are useful in following knock kneed children, and are easier to carry out than measurements of the tibiofemoral angle.

RADIOLOGICAL DEVELOPMENT OF THE FEMORAL EPIPHYSES

In the early weeks of fetal life, the femur is represented by a cartilaginous model, which resembles in form and shape the bone which will ultimately replace it. A centre of ossification appears in the shaft about eight weeks after fertilization of the ovum; growth in length is brought about by the extension of the growth centre upwards and downwards; at birth, a secondary growth centre, the epiphysis, appears at the lower end of the femur (*Figure* 6.6).

Serial radiograms of the knees from birth to the age of seven show that most of the increase in *width* of the knee joint which occurs is due to enlargement, torsion and alteration in shape of the medial condyle. In the early years, the medial condyle increases in transverse width, rather than in certical depth, and its vertical height does not exceed that of the lateral condyle until the child is six years old. For this reason, knock knee, occurring, as it frequently does, at the age of two, cannot be ascribed to 'growing down' of the medial condyle.

Age shape of Epiphysis		Disseminated calcification or ossification		Possible clinical features
		MC	LC	
Birth				
6 months	Oval			
1 year	Egg-shaped			
1½ years	Egg-shaped			In-toeing
2 years	More pointed medially; wooden shoe	+	0	Malleoli apart knock knee
2½ years	Wooden shoe	++	+	,,
3 years	Condyles equal in depth	++	+	,,
3½ years	,,	+++	+	Peak incidence knock knee
4 years	,,	++	+	Knock knee
5 years	,,	+	+	Malleoli nearer together
6 years 6 years	Increase in depth, medial condyle	+	0	Knock knee corrected.

LM

MC medial condyle LC Lateral condyle.

Figure 6.6. Development of femoral epiphyses (diagrammatic) (Petersen, 1967. Reproduced by permission of the publishers)

In ordinary x-ray films, only the ossified portions of the epiphyses can be seen; in x-rays taken to show soft tissues it is often possible to discern the faint outline of the cartilaginous model. Areas of disseminated calcification (or ossification) can be seen streaming out towards the periphery of the cartilage. When the borders of the condyles

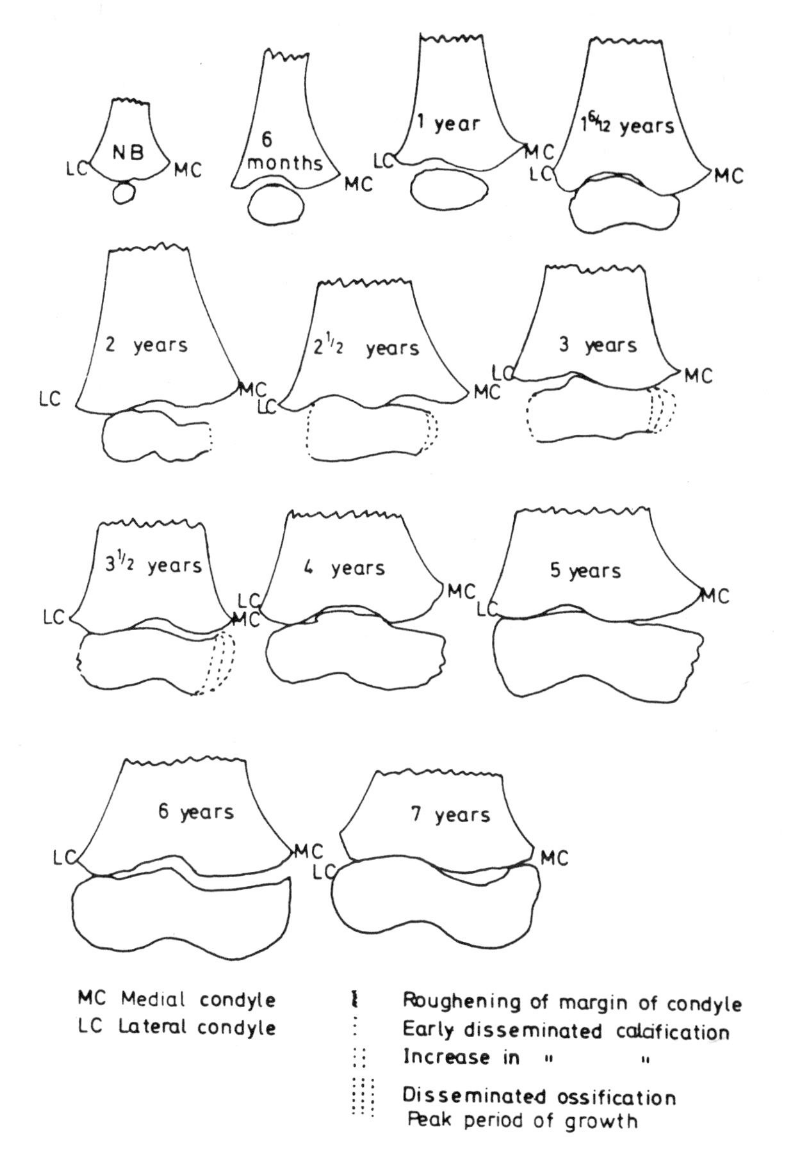

Figure 6.7. Tracings of femoral epiphyses from x-ray films from birth to seven years (Petersen, 1967. Reproduced by permission of the publishers)

appear smooth in the x-ray, ossification has probably reached the limit of the model.

Serial x-rays show that the epiphysis goes through characteristic changes in size and form, and that the pattern varies very little in different children; the ages at which changes are seen depends on stage of skeletal development. Children with developmental knock knee show the same radiological stages as those without knock knee.

Radiological changes in the lower femoral epiphysis in the first seven years

At birth the lower femoral epiphysis is ovoid in shape; at six months it is egg-shaped, and it retains this shape until the child is a year old. At eighteen months it becomes more pointed medially; its appearance at this stage has been compared to that of a wooden shoe. At two years the 'wooden shoe' is still apparent; the lateral condyle is larger in vertical height than the medial; disseminated calcification is beginning to be apparent at the medial margin of the epiphysis.

At 2½ years the wooden shoe is still apparent; disseminated calcification is marked in the medial condyle and less marked in the lateral condyle. At 3 years the wooden shoe is no longer apparent; the condyles are now equal in depth. Disseminated ossification is now seen in the medial condyle, and calcification is seen to a slight degree in the lateral condyle.

At 3½ years the condyles are equal in vertical depth; disseminated ossification has increased in the medial condyle. This is the peak age for incidence of disseminated ossification and also for knock knee.

At 4 years disseminated ossification is shown as a roughening of the medial border of the condyle; this is still present at the age of five. At 6 years the internal condyle begins to increase in depth; knock knee (if it has been present) will probably have been corrected. At 7 years, the medial condyle exceeds the lateral in vertical height as well as in transverse width; there is downward growth of both condyles.

Summary

The lateral condyle is larger than the medial in vertical height before the wooden shoe stage; compared with the medial condyle it grows very little in width. The medial however grows rapidly in transverse width, and by three years has nearly caught up the lateral condyle in vertical height. By the time the child is six years old, medial vertical height is greater than that of the lateral condyle.

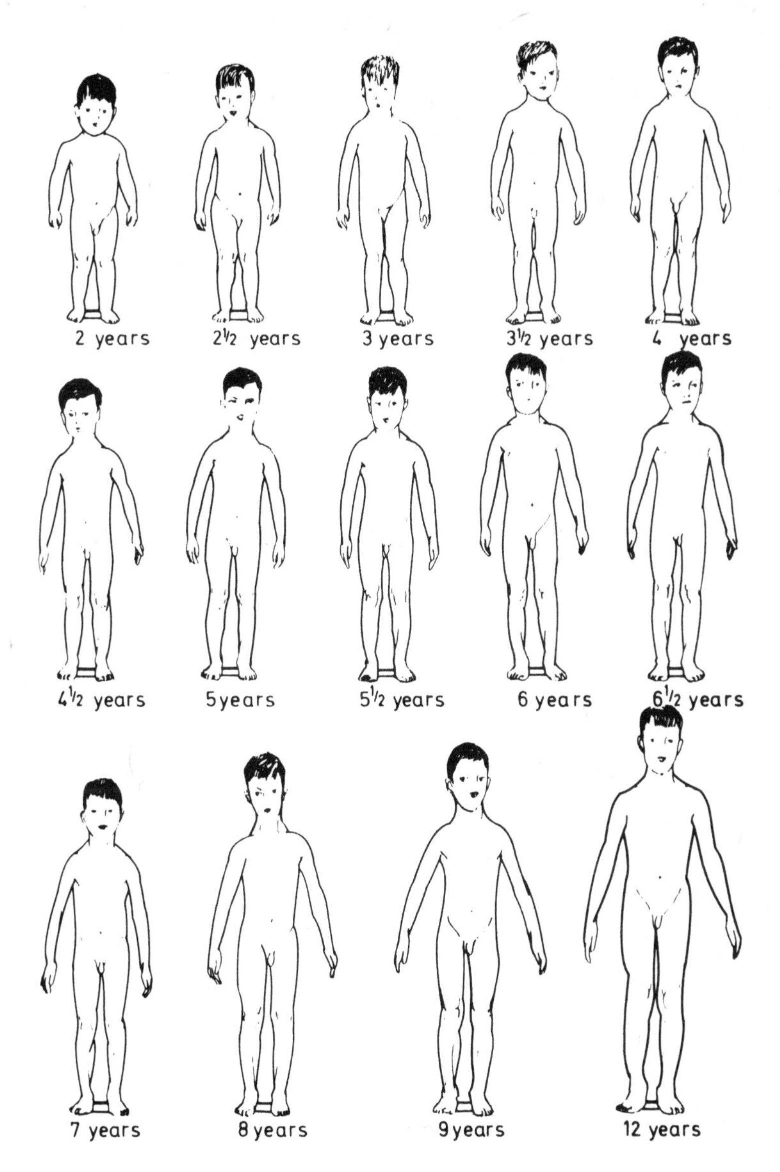

Figure 6.8. Development and correction of knock knee drawn from serial somatotype photographs of a boy aged from two to twelve years. The block separates the malleoli (Dupertuis and Tanner, 1950). Correction begins at the age of four, when medial condyles no longer touch.

LONG-TERM STUDIES

In *Figure* 6.8 a series of line drawings from one child is shown; these are traced from somatotype photographs. The child was photographed while standing on a turntable, while a T-shaped block kept his feet in position; thus, the internal malleoli are the same distance apart in all the drawings. The development and correction of knock knees can be followed if the spaces between the medial femoral condyles and between the thighs are studied; correction can also be observed by studying the curves of the lower end of the tibiae.

Knock knee is shown to be present from the age of 2 years onwards; at 4 years a gap is appearing between the medial femoral condyles; this is also seen at the age of four and a half. At five, the lower third of the tibia is beginning to show internal torsion; this is very well marked at the age of six when knock knee is usually corrected. After this age, growth of long bones is accelerated, and curve is less marked; most of these children show sharpness at the tip of the internal malleolus, which may persist until adolescence.

To summarize:

Age 2: Knock knee and slight external rotation of tibiae
Age 3½: Knock knee well marked
Age 5 onwards: Internal torsion of tibiae
Age 6: Knock knee corrected.

A summary of the results of the longitudinal studies is given below.

The sample consisted of 100 boys and 85 girls. Sixty-one of the boys and 44 of the girls had some degree of knock knee during infancy and childhood.

		Boys	Girls
(1)	Admitted under the age of 7½ Knock knee corrected by 7½	57	31
(2)	Admitted under 7½ Knock knee not corrected by 7½	5	5
(3)	Admitted under 7½ Knock knee corrects before 7½ but reappears	1	1
(4)	Admitted under 7½ without knock knee knock knee appeared for the first time under 7½	0	1

(5)	Admitted after 7½ with knock knee	0	6
(6)	Admitted after 7½ without knock knee; Knock knee developed later	7	0

In *Figures* 6.9 and 6.10 the duration of the knock knee and age of

Age years

1 2 3 4 5 6 7 8 9 10 11 12 13 14 15 16 17 18 19 20

		Fat Percentiles		
		7 years	10 years	15 years
1.	Admitted under the age of 7½; knock knee developed and was corrected by 7½ (47)			
2.	Admitted under 7½; knock knee present; not corrected by 7½ (5)			
151		70/60	97/97	
324				
336		45/55	35/45	25/5
426		25/25	50/35	35/25
533		25/35	15/5	
3.	Admitted under 7½; knock knee corrects before 7½ but re-appears (1)			
640		3/20	5/20	5/25
4.	Admitted after 7½ with knock knee (1)			
622			20/20	75/55
5.	Admitted after 7½ without knock knee; knock knee development later (7)			
227				30/15
236			70/25	50/25
249		85/95	90/97	
254			87/92	
257			75/90	
355		75/30	70/50	
354		75/30	70/50	

Knock knee absent ————————
Knock knee present – – – – – – – –

Figures on right of table correspond to fat percentiles; subscapular and triceps skinfold.

Peak age for growth shown thus ↑

Figure 6.9. Long-term studies of knock knee; boys

correction is given. The fat percentiles for each child are shown at the ages of 7, 10 and 15 years. Some clinical details of the children concerned are given.

To say that a child is in the tenth percentile for fat means that 10% of all normal children are thinner than that child. Children in the 90th percentile are too fat; the 50th percentile is average. Children outside the 10th to 90th percentile range should be kept under observation. To obtain fat percentiles the amount of fat is estimated by the skinfold method (i.e. Tanner 1962 at seven, ten and fifteen years of age. These readings, are compared with standards).

Notes on Figure 6.9. (Boys)

Case No.

151	This is a fat boy. Skinfold percentiles 97/97 at age of 10. Knock knee still present at 13.
336	Knock kneed until the age of 16; two other members of the family (M) (girls) failed to correct at 7½. Probably genetic in origin.
324	Knock kneed until he was 9 years old. Slow skeletal growth; changed from 15th to 25th percentile (height) at 9. Possibly correction of knock knee effected by growth in length and, perhaps, torsion of the femur.
426	Knock kneed until the age of 15 corrected spontaneously immediately after he had passed his peak in height growth.
533	Thin boy. Knock knee resolved at 10.
640	Knock knee appeared at 12; corrected at 14½. Height 50th percentile; not fat.
622	Late puberty. Knock knee corrected at 14½. Thin boy; height 10th percentile.
227	Knock knee corrected at 11½.
236	Knock knee corrected at 15; peak age (growth) 15.
254	Knock knee corrected at 11½; peak age (growth) 12½.
355)	
354)	Twins with identical pattern.
249	Fat boy; knock knee still present at 12½
257	Fat boy; knock knee still present at 13; (brother of 151).

Notes on Figure 6.10 (Girls)

Case No.

403	Fat girl of average height. Early menarche (12½) knock knees were corrected during the growth spurt.
514	Fat girl who failed to correct her knock knee at 7½; correction occurred during the growth spurt; early menarche.
318	This child belonged to one of the knock knee families (M). Classic developmental knock knee which resolved at 8 but reappeared when she put on weight at 13, and disappeared after menarche (14½).

539 This case has a genetic basis (family L); knock knee still present at 16.
627 This girl is also a member of family (L); knock knee still present at 16.
209 Knock knee reappeared at the age of 10, and was still present when she was seen at 20. Weight increased during growth spurt.

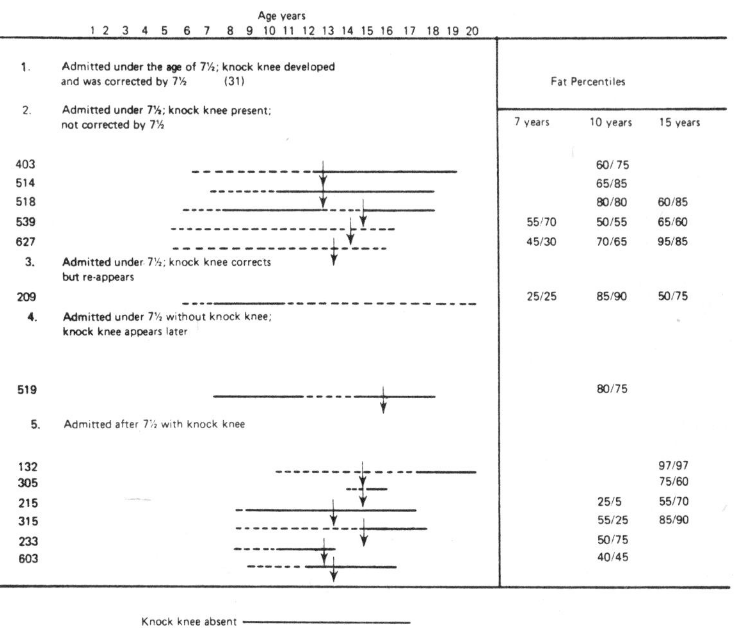

Figure 6.10. Long-term studies of knock knees; girls

519 Member of family (M). Knock knee short-lived (11–13).
132 Obese girl; knock knee corrected at 16.
305 Fat girl; knock knees corrected at 14½; menarche 14½.
315 Knock knee corrected at 14½; menarche at 14½.
215 Knock knee corrected at 8½; short girl; 10th height percentile.
233 Knock knee corrected at 9½.
603 Knock knee corrected at 11½.

CLINICAL SUMMARY

Development of knock knee

There are very few children in whom a mild degree of separation of the malleoli cannot be demonstrated and about 80 per cent of children show malleolar separation of one inch or more at some time or other. Unless special examinations are carried out these changes may pass unnoticed; they are slight, unimportant and soon corrected. The sequence of events described below applies to most children.

When the child begins to practice standing and walking in the second year he needs a broad base to help him in lateral balance; he obtains this by walking with his legs widely separated. His large head, long body, and short legs make him top-heavy, and he adjusts to gravity antero-posteriorly by tilting his pelvis; his abdomen may protrude and he may have lordosis. His femora, which were straight at birth, begin to bow slightly with a gentle concavity backwards.

As walking progresses the child may turn his toes in; if he walks early, that is, before he is a year old, this is probably due to internal femoral torsion. If he begins to walk after the age of fifteen months the in-toeing may be the result of internal torsion of the tibiae, and may be a device to help the child with his balancing problems. As growth in height proceeds the knees are brought nearer together; this change is probably due to the change in direction of the femur, or possibly to femoral torsion. The child, however, still needs a large base for lateral balance, and, at this stage, the lower ends of the tibiae may rotate externally, thus creating a gap between the malleoli; this is the beginning of 'developmental knock knee'. By the time the child is two, the malleoli may be an inch or more apart. Meanwhile the femoral condyles have increased in width (*see* page 63), the medial condyle rotating and increasing in width faster than the lateral and therefore protruding medially (but not downwards) and pushing the internal malleoli further apart. Thus there are several contributory causes of knock knee in the early years. The incidence of knock knee is at its highest in the period 3–3½ years, and 3½ years is the peak age for disseminated calcification or ossification (*see Figures* 6.7, 6.8). At this age there may be malleolar separation of two or more inches.

Correction of knock knee

Children under 8 years of age

The distance between the internal malleoli is usually found to decrease after the child has arrived at the age of 3½ years. Serial photographs of children show that during the fourth and fifth years, the lower ends of the tibiae curve inwards and in due course the internal malleoli approach each other, and, in most cases, meet (*Figure* 6.10). This is probably effected by internal torsion of the tibiae. Correction can be expected to be complete between the ages of 5 and 7 years. Two other factors assist in correction between the ages of 5 and 7; the femur begins to increase in length faster than the tibiae after the age of 5, and the pad of fat has usually disappeared from the inner aspect of the internal condyle. Whatever the mechanism, most cases of knock knee correct spontaneously before the age of seven.

Children of 8 years and over

If knock knee has not resolved by the means described above, other mechanisms may promote correction. Increase in hip width at the time of the adolescent spurt may have the effect of drawing the femoral heads apart, so that the medial femoral condyles are separated from each other. Increase in length of long bones and possibly alteration in femoral torsion may assist in correction of knock knee. This form of correction is commoner in boys than in girls.

Finally, knock knee which is still present at 15 or thereabouts may resolve at the end of puberty after menarche in girls, and after peak in height in boys.

Causes of persistence of knock knee after age of 7½ years

In some cases, correction is merely a little delayed and may occur by 8 or 9. This is most frequently the case in children who are under average height and skeletal age (*see Figures* 6.9, 6.10).

Many children over 7 years of age with knock knee are overweight; some of them have always been fat and have not been able to correct their developmental knock knee by internal torsion of the tibia. Part of their difficulty is due to the presence of fat pads separating the internal condyles. These children, like fat toddlers, have a balancing problem and need a wide walking base; they may be relying for this on external torsion of the tibiae. The pelvic tilt is not usually increased; lateral balance seems to be the main difficulty.

There appears to be a genetic basis in some of these cases; in two of the families studied (L and M) the tendency to knock knee is marked. X-rays may show a characteristic deviation in the upper third of the shaft of the tibia. Other cases are due to alteration in femoral torsion. This type of case rarely improves during growth spurt, and if the knock knee is severe it may warrant orthopaedic treatment.

SUMMARY

(1) Knock knee is defined as a condition in which the medial malleoli fail to meet when the child sits on the floor with his legs stretched out in front of him.

(2) Knock knee is present in some degree at some stage of development in about 80 per cent of children, and its occurrence may be considered as a stage of normal development.

(3) Knock knee is usually seen first at the age of 2 years. The peak incidence is at the age of 3½ years; between the ages of 5 and 7 correction usually takes place.

(4) At 2 years the lower end of the femur enlarges; most of this increase in size is due to the widening of the medial condyle. Femoral torsion has decreased since birth and evidence of recent changes in position of intercondyloid fossa can be seen in the popliteal space.

(5) Different factors, such as fat deposits, and variations in rate of growth, may delay the correction of developmental knock knee. In such cases, correction may be completed during the growth spurt, either at its inception or after the menarche or peak of growth in height in boys.

(6) When knock knee persists after the growth spurt is over it may be genetic in origin.

REFERENCES

Appleton, A. B. (1964). 'Posture'. *Practitioner,* **156,** 5

Dupertuis, C. W. and Tanner, J. M. (1950). 'The pose of the subject for photogrammetric anthropometry with special reference to somatotyping.' *Am, J. Phys. Anthropol.* **8,** No.1., 27

Frazer, J. E. S. (1965). *'Anatomy of Human Skeleton.* London: Churchill

Morley, A. J. M. (1957). 'Knock knee in childhood.' *Br. med. J.* **2,** 976

Petersen, G. (1967). *Atlas for Somatotyping Children.* Netherlands: Van Gorcum

Roberts, W. H. (1962). 'Femoral torsion in normal human development as related to dysplasia.' *Anat. Rec.,* **143,** 369

Additional reading:

Pyle, S. I., and Hoerr, N. L., (1955). *Radiolographic Atlas of Skeletal Development of the knee.* Springfield, Illinois: Thomas

Tanner, J. M. (1962). *Growth at Adolescence,* 2nd edn. Oxford: Blackwell

CHAPTER 7

7 Bow Legs

PHYSIOLOGICAL

Bow legs are common in infancy and are of little clinical importance. They rarely need treatment although the parents often need to be reassured that their children's legs will straighten themselves in due course.

When the child begins to walk he does not always find that balancing comes naturally; he has to practise in order to find out how to stand erect without falling; this, he finds, is more easily achieved if the legs are held further apart, so helping in lateral or side-to-side balance. When his feet are wide apart the napkins tend to fill out the space between the legs and this wet bundle of napkins makes the child look far more bow legged than he really is. If he walks at the age of one year the bowing of the legs may be noticeable and is due to incurving of tibiae. As time goes on the tibiae rotate and turn outwards, thus gradually straightening the legs; this may happen in the second year. However, the legs do not remain straight for very long and the tibiae soon turn outwards and the internal condyles meet. Knock knees may follow (*see* Chapter 6). (*Figure* 7.1 shows the curvatures which are seen, and torsions of the bones involved. The curves are compounded of external femoral torsion, internal femoral and lateral tibial curvature. When child first stands the straight femur begins to show a gentle anterior-posterior bowing (convex anteriorly). The internal femoral curves next appear and toes may turn in. Bowing is now apparent but is soon replaced by medial rotation of the tibiae. The child is beginning to 'undo' the bowing preparatory to straightening his legs. However knock knee soon begins to be apparent (*see* Chapter 6).

Note. This sequence is probably correct, but other interpretations may, of course, be put on these appearances;

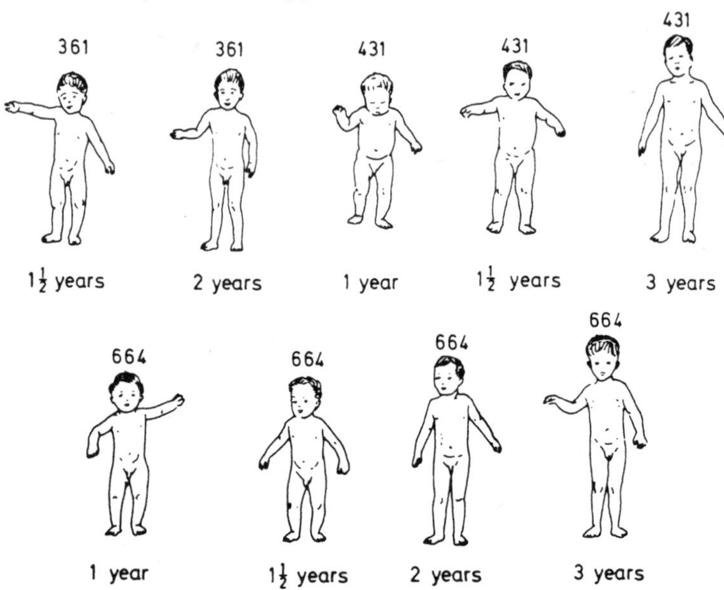

Figure 7.1. Bow legs; three children in age group 1–3 years

No. 361	*Age 1½ years–Bowing well marked*
	Age 2 years–Tibiae straightening
No. 431	*Age 1 year–Tibiae straightening*
	Age 1½ years–Tibiae straightening
	Age 3 years–Knock knee developing
No. 664	*Age 1 year–Early bowing tibiae*
	Age 1½ years–Bowing of femur (anterior-posterior
	Age 2½ years–Tibiae straightening
	Age 3 years–Knock knee

METABOLIC
(BOWING OF THE LEGS ASSOCIATED WITH RICKETS)

Active rickets is now a comparatively rare disease in Western countries. 'In white children severe *nutritional* rickets is extremely rare, but an occasional case of vitamin-D-resistant rickets is found in older children or it may complicate some renal diseases or malabsorption states. All forms of rickets can affect the tibiae when characteristic bowing may be seen, i.e. the tibiae bend backwards and inwards.

Rachitic bowing may not be noticed until after walking has started. Formerly, many rachitic bow legs were treated surgically, and osteoclasis appeared regularly on the hospital operation lists. (It was unpleasant to hear the sound of the children's legs being deliberately fractured in this way, though the results were good.) Children were given massive doses of vitamin-D, and were generally discharged with straight serviceable limbs.

BLOUNT'S DISEASE (PRIMARY TIBIA VARUM)

If only one leg is bowed in primary tibia varum it is possible that the condition is structural; bowing may however be present in both legs and is then known as Blount's disease. The appearance is a combination of lateral curvature and internal rotation. It is said to be common in West Indian children; they walk early, and this may be one of the causes of the disorder.

Blount's disease is progressive and non-correcting. The x-ray shows irregularity of medial tibial metaphysis and flattened epiphysis. The disease is treated by tibial osteotomy and if changes have not been too severe a good result can be expected.

CHAPTER 8

8 Flat Foot and Valgus Heels

In this chapter the anatomy of the region is briefly reviewed, and the results of a longitudinal study of posture in children are discussed.

Method of study

185 children were examined at 6-monthly intervals, during exercise and at rest. Footprints were recorded and foot lengths were measured.

ANATOMY

Man is an erect animal and his feet must provide not only leverage when walking and running, but a solid base on which the body can be balanced during standing. He has acquired such feet by the process of natural selection; modification of the tarsus and metatarsus have produced the bony arches and the valgus mechanism of the heel. The arches make possible leverage on running and stability on standing; the valgus mechanism provides for inversion and eversion of the feet.

The arches of the foot (*Figure* 8.1)

Opinions regarding the number of longitudinal arches in the foot have differed. Some recognized two longitudinal arches, an outer and an inner; others described *one* longitudinal arch, having *one* component in the hind foot, and *two* components in the forefoot. The second view is the one most generally accepted.

The longitudinal arch consists of bony segments. The talus is the keystone; the calcaneus may be regarded as the posterior pier, and the heads of the first and fifth metatarsals as anterior piers. Thus, the foot may be likened to a tripod with three weight-bearing areas; heel, outer border of sole and ball of foot. The outer side of the foot, which is of slender build, is concerned with stability, balance and support; while the inner side, which is more massive, is concerned with weight-bearing

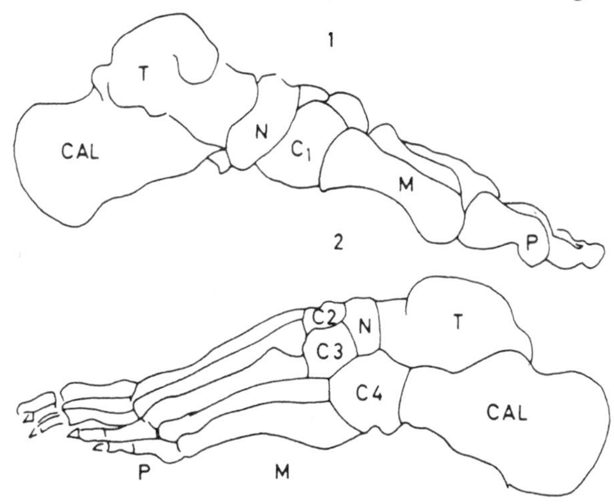

Figure 8.1. Arches of the foot

1. Medial view
CAL–Calcaneus
T–Talus
N–Navicular
CU–Cuboid

2. Lateral view
C1, C2, C3–First, second and third cuneiforms
M–Metatarsals
P–Phalanges

and propulsion. Ligaments, fascial planes and plantar muscles tie the pillars of the longitudinal arches together. The tibialis posterior and the peroneal muscles act as muscular slings under the apex of the arch; these muscles contract when the foot changes from a solid passive foundation to an active working member, and convert the arch to a fixed lever.

The longitudinal arch

There have been many arguments and discussions as to the relative importance of muscles and ligaments in maintaining the longitudinal arch. The generally accepted view is that the bony arch is maintained by ligaments; the anterior group of muscles does not normally hold up the arch. Electromyographic studies suggest (Joseph, 1960) that in normal standing very little muscular contraction is involved; the head and body are balanced on the lower limbs and feet, and there is little, if any, tension in the thigh muscles. Slight intermittent contractions of the soleus muscle may occur to prevent forward tilting towards the centre of gravity.

The transverse arch

The only transverse arch now recognized is the half dome; this is half an arch. When the two feet are parallel and in apposition, the dome is complete; the arch runs from cuboid to cuboid through the base of the metatarsals. In normal stance the head of each metatarsal bone bears part of the body weight; the first metatarsal bears twice as much weight as each of the others.

Ankle complex

The talus is the kingpin in the ankle complex; it has three articular surfaces, and takes part in the formation not only of the ankle joint proper, but of the subtalar and midtarsal joints.

The ankle joint proper

This is the articulation between the talus and the lower ends of the tibia and fibula; the talus has a smooth pulley-like surface slightly broader anteriorly than posteriorly. Dorsal and plantar flexion are the only movements that can be performed at this joint; plantar flexion from a right angle is possible up to an angle of 50 degrees, whereas dorsiflexion is only possible within a range of 15 degrees. When dorsiflexion takes place, the talus moves backwards, so that the broader part of the pulley comes to lie between the malleoli; the grip becomes stronger as dorsiflexion nears completion, and finally locking occurs. Conversely, in plantar flexion the narrower portion of the talus moves forwards; the grip is looser, and thus a certain amount of lateral mobility is possible.

The subtalar joint (Figures 8.2, 8.3)

This is the joint between the plantar surface of the head of the talus and the calcaneus; the calcaneus contributes three articular surfaces to the joint so that it is possible for the talus to rotate on an oblique axis by means of swivelling movements from side to side. Inversion and eversion take place at this joint; inversion occurs when the soles are turned towards each other, as if they were trying to imitate hand-clapping; and the foot is said to be everted when the sole is turned outwards, and the outer border of the sole is raised.

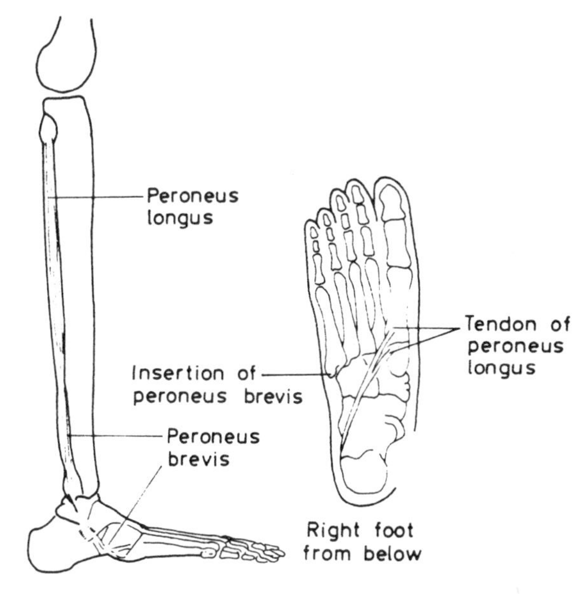

Figure 8.2. Eversion of the feet

The tibialis posterior is the most powerful of the invertors; its tendon passes under the internal malleolus and is inserted mainly into the navicular bone; it sends slips to most of the tarsal bones. The peroneal muscles are mainly responsible for eversion, which is well developed in the human species.

In the condition known as valgus ankle (or, more correctly, as valgus heel), the calcaneus rolls over into eversion, thus ensuring that all weight-bearing areas of the tripod are planted firmly on the ground. This balancing mechanism is of particular value in children with high arches, and should not be regarded as a defect. The invertors and evertors also enable the sole to adapt to plane surfaces; for example when walking round a mountain it is possible to invert one foot and evert the other.

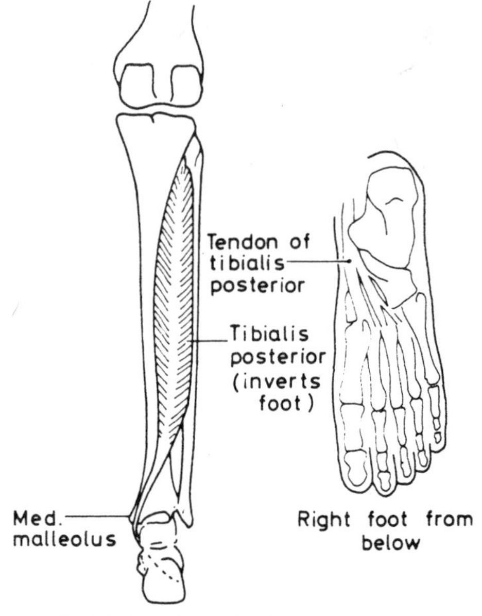

Figure 8.3. Inversion of the feet

The midtarsal joint

At this joint the talus articulates with the navicular and the calcaneus with the cuboid; there is very little mobility, though gliding movements are possible, and inversion at this joint may accompany eversion at the subtalar joint, causing pronation of the forefoot.

Anatomy of 'flat foot'

A child's foot may appear to be flat because he has a low arch, because he shows some degree of 'valgus heel', or because the longitudinal arch is obscured by a pad of fat.

Shape of arch

In general, the shape of the longitudinal arch varies very little in childhood; the pedograph reveals little change in the forefoot in serial footprints. The only dramatic change which may occur is seen in the infant, when an arch emerges from a 'flat foot' as the pad of fat disappears.

Throughout this book, 'flat foot' and low arches have *not* been regarded as synonymous; the most marked cases of clinical flat foot are usually due to the valgus position of heels with high or medium arches.

'Valgus heel'

The ability to roll the heel over in eversion in response to balancing problems is particularly useful in childhood and adolescence as constant readjustments have to be made to meet the demands of gravity.

These variations are clearly shown in the pedograph; if severe degree of valgus is present the whole of 'plateau space' may be obliterated. This appearance may be seen at adolescence, and is usually transient.

Rating of valgus heel on posture card.

(1) Mild valgus. Slight eversion at subtalar joint.
(2) Severe valgus. Eversion marked.
(3) Moderate to intermediate.

THE PEDOGRAPH FOOTPRINTS

Footprints were recorded six-monthly, using Scholl's pedograph; a normal footprint of an eleven-year-old child is shown on page 82. The tracing shows a forefoot, a hindfoot and a connecting bridge. The weight-bearing areas are marked W1, W2, W3; they represent the bases of the tripod (*Figure* 8.4). Skimming the part of the print representing the inner margin of the hindfoot, a line XY is drawn; this encloses a plateau-shaped space XYZ. A perpendicular PT is drawn from XY to the highest part of the plateau and a line AB across the narrowest part of the bridge. The method of rating 'arch' status is shown; this normal foot would be graded as 4.

The dotted lines numbered 1 to 7 correspond with rating on posture card (*Figure* 2.4). 'Medium' is rated as 4; 1, 2, 3 denote varying degrees of flatness, and 5, 6, 7 show degrees of height of arch. In this print, rated medium 4, ratio of width of bridge (AB) to plateau height (PT) is 1/3. AB and PT are measured to nearest half centimetre. These measures are only approximate but it is useful to be able to note the changes which occur in ratio AB/PT over the years in individual cases.

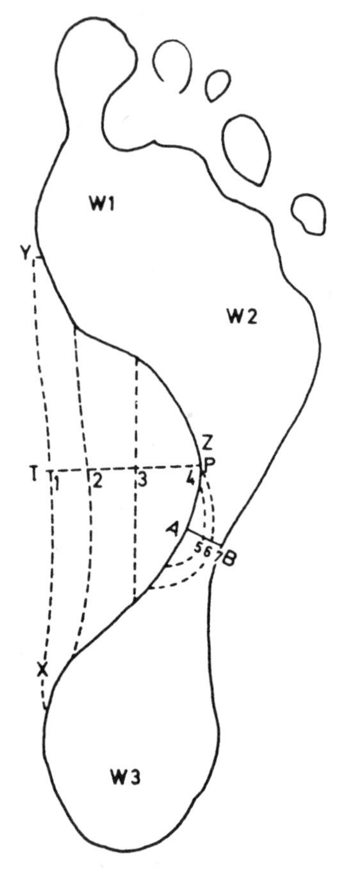

Figure 8.4. Footprint of normal eleven-year-old boy

Isolated prints do not yield information of much value, but serial footprints are well worth studying. The constancy of the arch pattern throughout growth and development can be watched, and the appearance and disappearance of valgus heel can also be observed. Summaries of a few representative cases are given below, and tracings of pedographs are reproduced.

What to look for in a footprint

(1) Ratio of bridge to plateau AB/PT

(2) Hind foot. The print of the hindfoot in the median child is pear-shaped; in early stages of valgus heel no alteration in shape may be seen, but a slight angulation at X may be seen as valgus increases. When valgus is pronounced, the plateau space may be encroached on, and may even be obliterated.

GH (534) M (*Figure* 8.5)

This child was first seen at the age of 2½. He had a thin foot and the

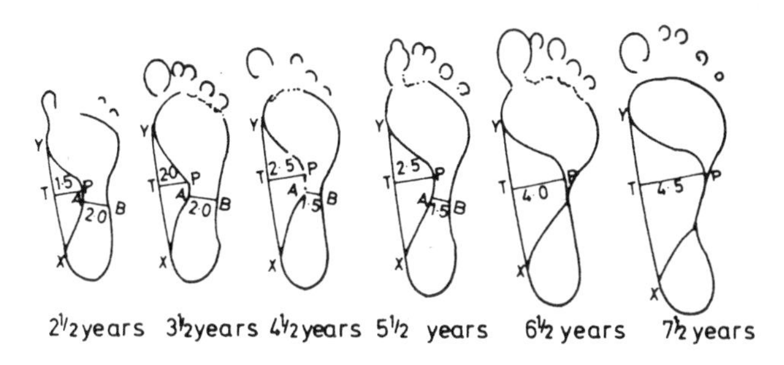

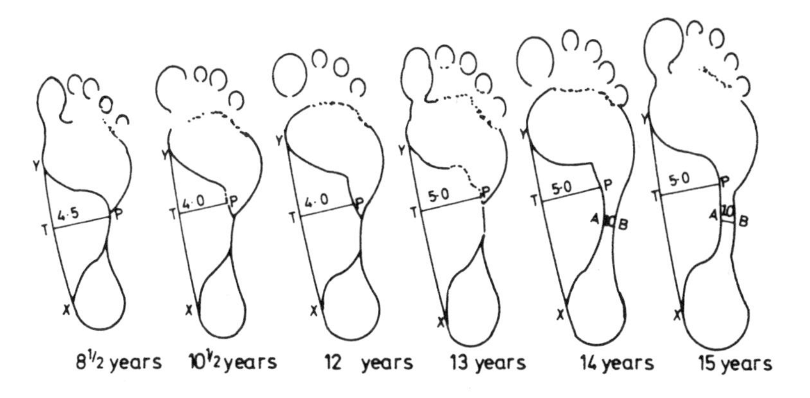

Figures 8.5. Serial footprints; GH (534)M

longitudinal arch was well developed; the bridge (AB) however was longer than the plateau height. At 3½ AB and PT were equal, but from 4½ onwards the plateau increased and bridge decreased, till at 8½ it is represented in the footprint by an attenuated line in print. Very little change is seen after this age. The clinical records show that from early days he had a mild valgus heel; from the age of 10½ onwards angulation at X shows that valgus heel had increased in severity.

The child's arch was visible at an early age; his initial flat-footed period was over before he was seen at 2½ years. The bridge: plateau ratio decreases over the years; at the age of 3½ the ratio was 1:1 The bridge became narrower as the plateau became higher.

Again and again, we find that children with long thin feet and high arches have recourse to valgus heels in order to maintain balance.

Age	2½	3½	4½	5½	6½	7½	8½	9½	10½	11½	12
AB	2	2	1·5	1:5	0	0	0	–	0	–	0
PT	1·5	2	2·5	2·5	4	4·5	4·5	–	4·0	–	4·0

TJ (661) M (*Figure* 8.6)

At the age of 6½, this child's longitudinal arch was regarded as normal but some angulation was present at X (characteristic of valgus heel) and the inkprint representing the inner border of the hindfoot was prolonged

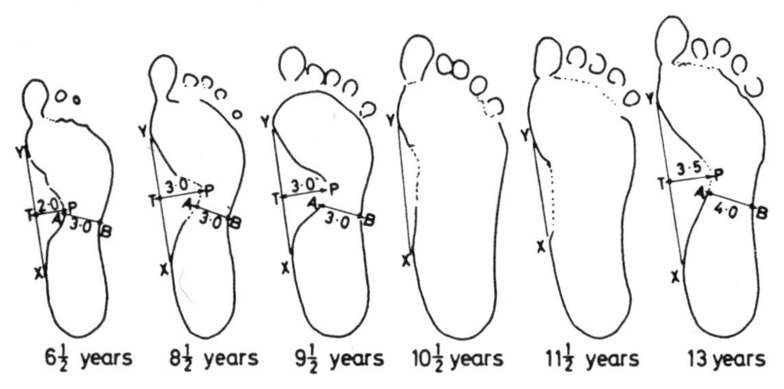

Figure 8.6. Serial footprints; TJ (661)M

forwards; valgus was also noted on clinical examination. By the age of 10½ the plateau space had been engulfed by soft tissues associated with severe valgus heel; the longitudinal arch could no longer be seen. By twelve, however, the arch was again visible.

Age	6½	8½	9½	10½	11½	12	12½	13½	14
AB	3	3	3	engulfed		3	4	5·5	5·5
PT	2	3	3	0	0	1	3·5	1·5	1·5

RT (621) M (*Figure* 8.7)

This boy had a medium to low arch, which retained its shape from the age of 8½ to 17, though the valgus heel mechanism partly obliterated plateau space during adolescent spurt (age 12) and again at the age of 15. At 15 part of the obliteration of the arch was due to increase in soft tissues, and part to eversion of the heel. At 17 the arch reappeared (AB/PT equals 3·5/3·5). This figure approximates to his nine-year-old ratio, and will probably remain with him during adult life.

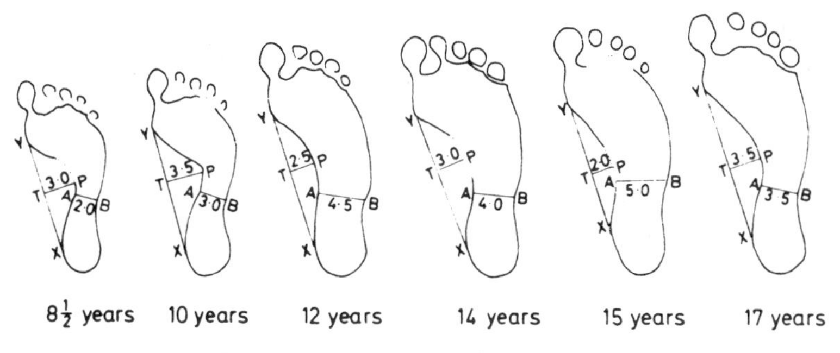

Figure 8.7. Serial footprints; RT (621)M

Age	8½	9	10	11	12	13	14	15	17
AB	2·0	3	3	3·5	4·5	3	3	5	3·5
PT	3	3	3·5	3	2·5	3	4	2	3·5

Peak height age: 13
Fat Percentiles Age 10: 35/40 (subscapular and
Age 15: 15/55 triceps skinfold)

LG (128) M (*Figure* 8.8)

This boy was first seen at the age of 3½. He showed the clinical picture of 'flat foot'; the inner border of the foot was straight, the arch was obscured by a pad of fat. At 4½ the longitudinal arch was well marked; the bridge measurement AB had decreased from 3·5 to 1·5, and by 5½ the arch was so high that no recording of AB or PT was possible (rating on posture card 6).

The *pattern* of the arch did not change; valgus heel was clinically present during the peak period of growth. It is represented in the pedograph as a slight angulation in the pear-shaped print of the hind-foot. By the age of 11½ an attenuated bridge made measurement of PT possible.

Valgus heel is very common in children with high arches and is part of a balancing mechanism.

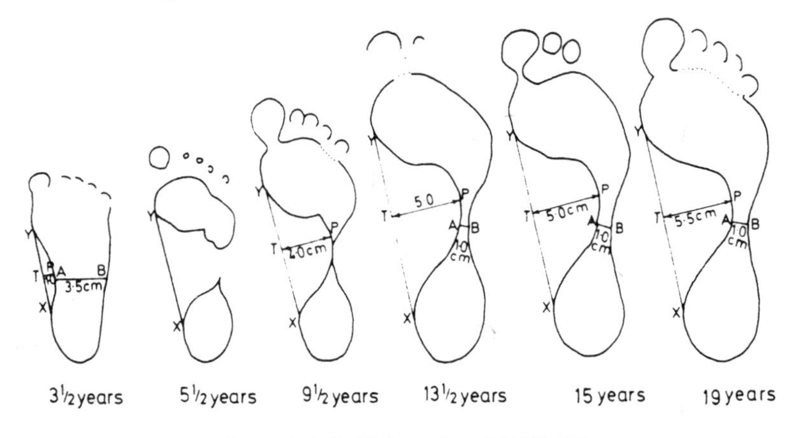

Figure 8.8. Serial footprints; LG (128)M

Age	3½	4½	5½	9½	11½	13½	15	17
AB	3·5	1·5	0	0	0·5	1·0	1·0	0·5
PT	1·0	2·5	0	4·0	4·0	5·0	5·0	5·0

LB (120) F (*Figure* 8.9)

This child had an infantile flat foot at the age of 2: at 6 years she developed a high arch, and retained her infantile valgus to enable her to cope with it. Angulation is seen (X) up to the age of 12. Valgus heel is apparent in several of the photographs (*Figure* 8.9) where irregularity of the inner border of the foot below the medial malleolus is seen; this is possibly due to displacement of the navicular tubercle and partial separation of the talonavicular joint. This is particularly well shown at age 11·5.

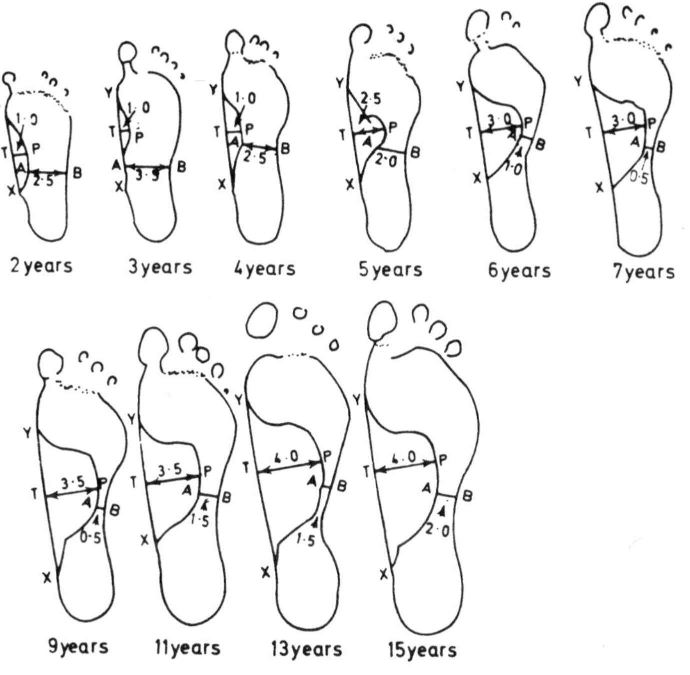

Figure 8.9. Serial footprints; LB (120)F

The bridge: plateau ratios are given below.

Age	2	3	4	5	6	7	8	9	10	11	12	13	15
AB	2·5	3·5	2·5	2·5	1·0	0·5	0·5	0·5	1·0	1·5	1·0	1·5	2·0
PT	1·0	1·0	1·0	2·5	3	3	3·5	3·5	4	3·5	4	4	4

LB (120) F (*Figure* 8.10)

These photographs of legs were enlarged from serial somatotype photographs. At the age of 11½ the irregularity in outline due to prominence of the navicular is apparent.

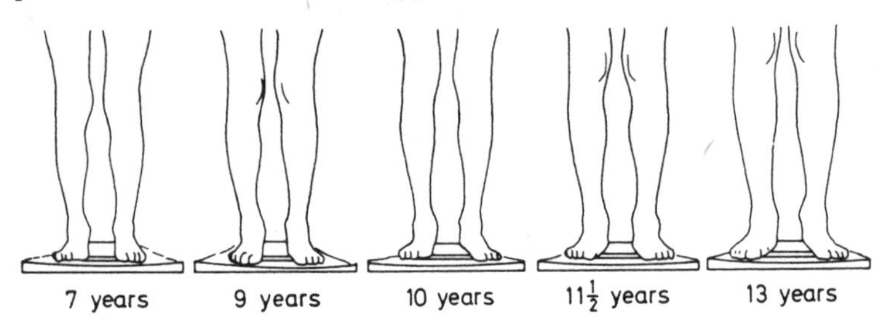

Figure 8.10. Enlargement from somatotype photographs of legs of child showing prominence of the navicular tuberosity; (LB(120)F)

FOOT LENGTH

Foot length was estimated six-monthly, using a foot rule with a sliding stop. *Figures* 8.11 and 8.12 show the median curve and percentiles for foot length for both sexes.

These curves show marked resemblance to curves for growth in height; there is the same sex difference, i.e. earlier growth spurt, and earlier peak height age in girls. It is interesting that the foot length of girls is always less than that of boys of same chronological age. It must be remembered, however, that the foot length represents length of metatarsals and phalanges as well as the length of tarsal bones. It is possible that the tarsal and metatarsal bones have different growth patterns, and that metatarsals follow the main pattern for long bones, and the tarsal bones may grow at quite different rates. It appears, that, in boys, the peak age of growth of the foot is earlier than that of general skeletal growth. In girls the curves for foot length and for general growth in height appear to resemble each other closely after the age of fifteen.

CROSS-SECTIONAL STUDIES: LONGITUDINAL ARCHES AND VALGUS HEELS

The number of children studied varied at each age. In general the samples of children under 5 and over 15 years old were not large enough to give significant results. Longitudinal arches (*Figures* 8.13, 8.14).

In both sexes, the percentage of 'flat' footprints was high in the pre-school child, but, by the age of 6½, the percentage had decreased. This was not due to an alteration in the arch, but to loss of fat and to development of postural reflexes.

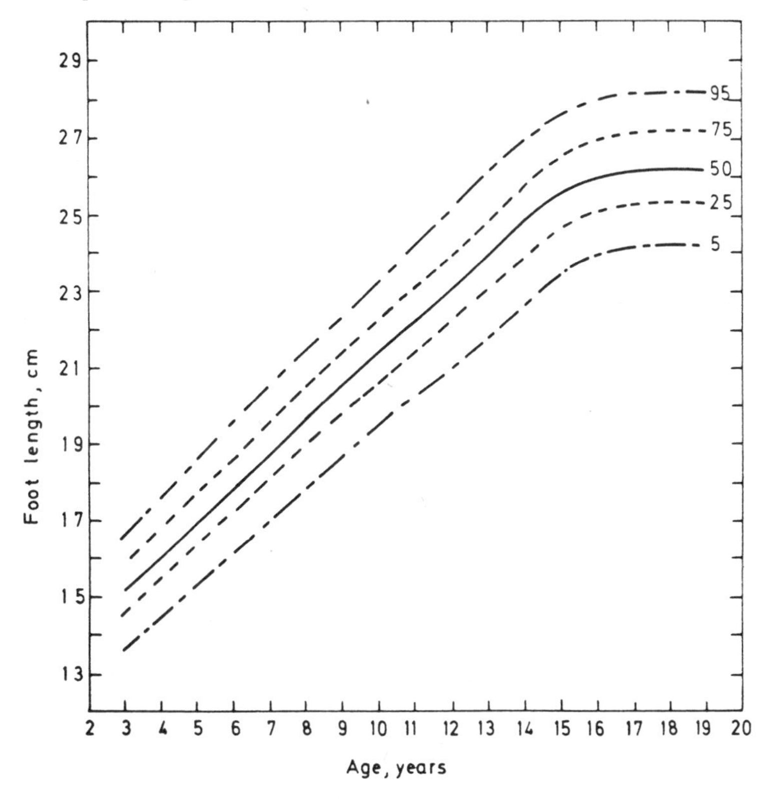

Figure 8.11. Growth of foot in length; boys

Valgus heels (*Figures* 8.15, 8.16)

In both boys and girls under the age of 5 years, 70 per cent showed some degree of valgus at some time. In *Figures* 8.15, 8.16, mild, severe and intermediate types of valgus are plotted separately, and it is apparent that between the ages of 9½ and 11 (girls) and 13 and 15 (boys) there is a rise of incidence of Grade 2 valgus, probably associated with the adolescent spurt.

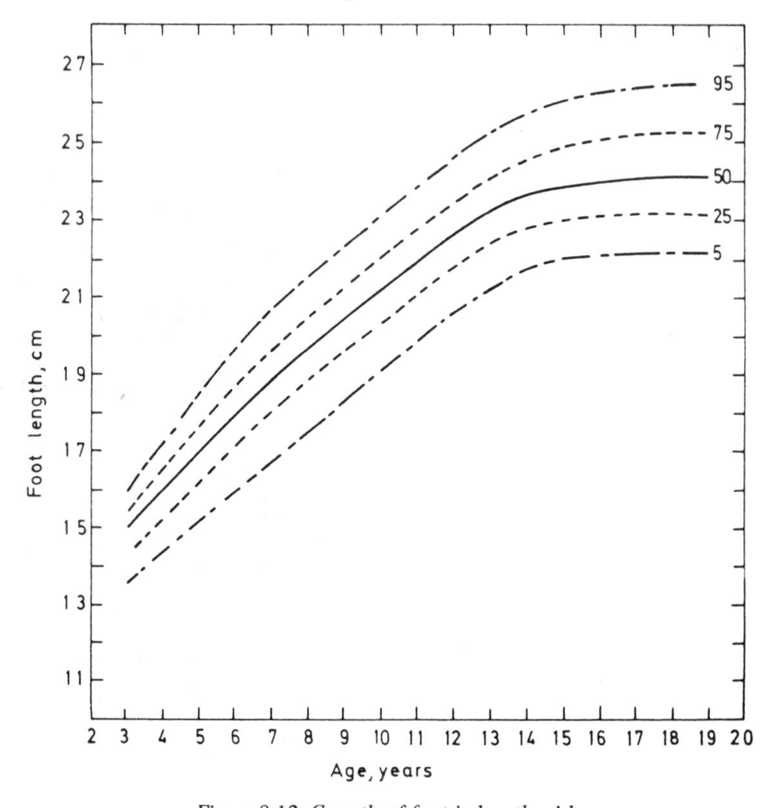

Figure 8.12. Growth of foot in length; girls

These transverse studies show certain trends but they are not so informative as the longitudinal studies given below.

LONG-TERM STUDIES OF LONGITUDINAL ARCHES OF FOOT AND VALGUS HEELS (100 BOYS AND 85 GIRLS)

Children's feet were inspected six-monthly during standing, walking and running, and in the long-sitting position on the floor. A recording of the footprint was examined.

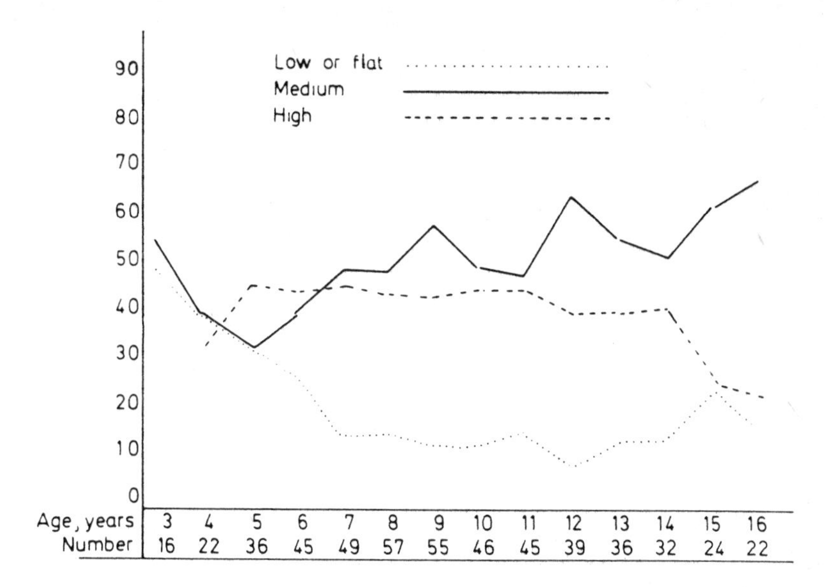

Figure 8.13. Cross-sectional studies of long arches; boys

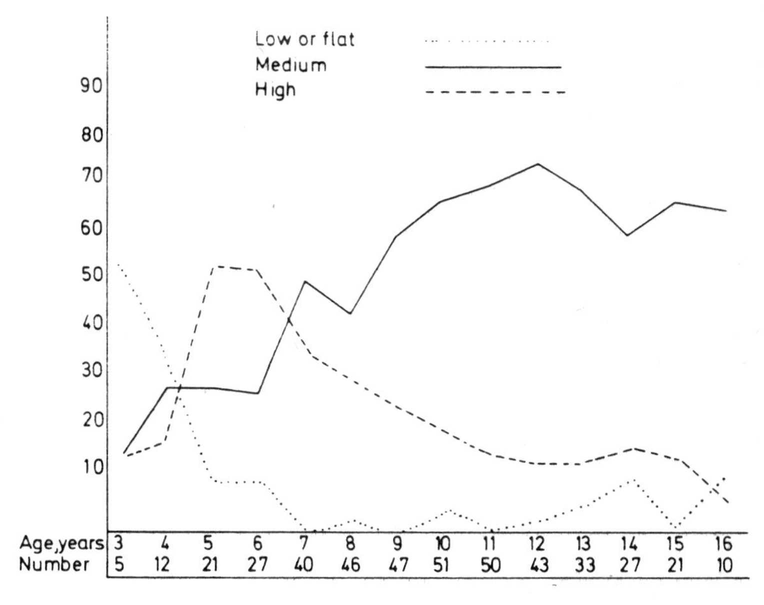

Figure 8.14. Cross-sectional studies of long arches; girls

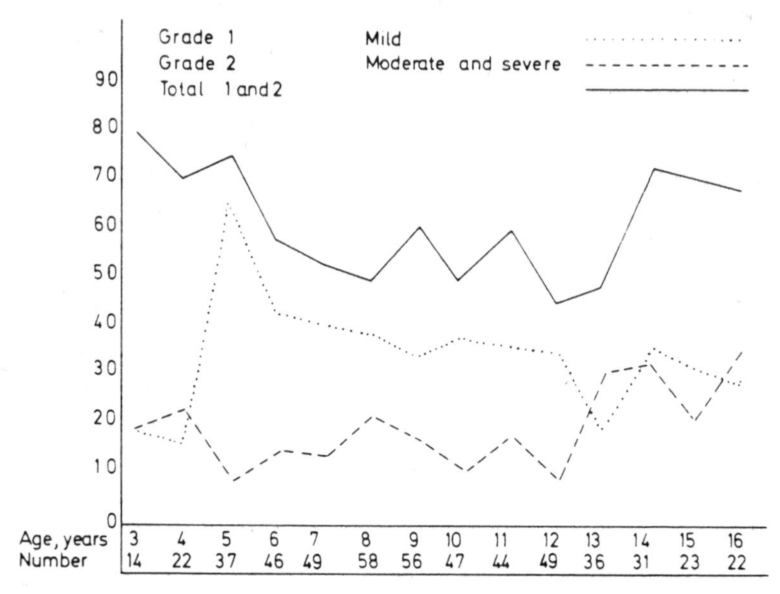

Figure 8.15. Cross-sectional studies of valgus heels; boys

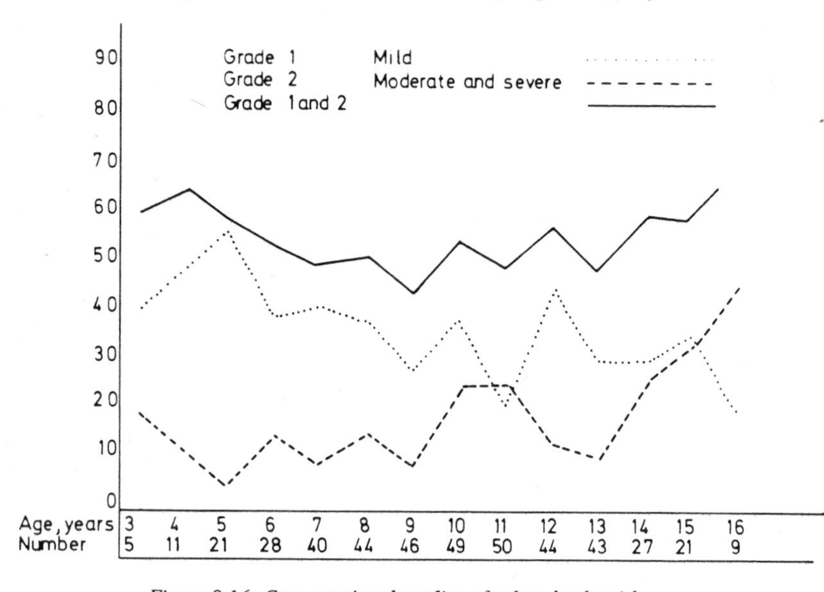

Figure 8.16. Cross-sectional studies of valgus heels; girls

LONGITUDINAL ARCHES OF FOOT AND VALGUS HEELS

Table 8.1 shows in how many children the arch and valgus patterns remained constant during the period of study and in how many the patterns changed.

TABLE 8.1

	Boys	*Girls*
Number of children with arch pattern constant	73%	75 (88%)
Number with valgus pattern constant	79%	58 (68%)
Number with arch pattern changed	27%	10 (12%)
Number with valgus pattern changed	21%	27 (32%)
Number with both arch and valgus pattern changed	7%	5 (6%)

The following two-way Tables are also given:

TABLE 8.2

Arches of Foot and Valgus Heels: 100 Boys

		Valgus heels				
Long arches	*Total* %	*0*	*1*	*2*	*3*	*Variable*
Medium arches	29	0	12	16	0	1
Flat Arches	20	1	6	13	0	0
High Arches	25	5	8	7	0	5
Variable	26	0	5	16	1	4
Total	100	6	31	52	1	10

TABLE 8.3

Arches of Foot and Valgus Heels: 85 gm.

		Valgus heels				
Long arches	*Total* %	*0* %	*1* %	*2* %	*3* %	*Variable* %
Medium Arches	44 (51·8)	1 (1·2)	18 (21·2)	17 (20·0)	2 (2·4)	6 (7·2)
Flat Arches	11 (13·2)	1 (1·2)	4 (4·8)	4 (4·8)	2 (2·4)	0 (0)
High Arches	21 (24·7)	3 (3·6)	6 (7·06)	10 (12·0)	0 (0)	2 (2·4)
Variable	9 (10·6)	1 (1·2)	2 (2·4)	2 (2·4)	1 (1·2)	3 (3·6)
Total	85 (100)	6 (7·2)	30 (36·0)	33 (38·8)	5 (6·0)	11 (13·2)

TABLE 8.4

Valgus Heels and Arches of Foot; 100 Boys

		Arches			
Valgus Heels	*Total*	*Medium*	*Low*	*High*	*Variable*
Valgus heel 0	6	0	1	5	0
Valgus heel 1	31	12	6	8	5
Valgus heel 2	52	16	13	7	16
Valgus heel 3	1	0	0	0	1
Variable	10	1	0	5	4
Total	100	29	20	25	26

TABLE 8.5

Valgus Heels and Arches of Foot; 100 Boys

			Arches		
	Total %	*Medium* %	*Low* %	*High* %	*Variable* %
Valgus heel 0	6 (7·06)	1 (1·2)	1 (1·2)	3 (3·6)	1 (1·2)
Valgus heel 1	30 (36·0)	18 (21·2)	4 (4·8)	6 (7·2)	2 (2·4)
Valgus heel 2	33 (39·6)	17 (20·0)	4 (4·8)	10 (12·0)	2 (2·4)
Valgus heel 3	5 (6·0)	2 (2·4)	2 (2·4)	0 (0)	1 (1·2)
Variable	11 (13·2)	6 (7·06)	0 (0)	2 (2·4)	3 (3·6)
Total	85 (100)	44 (51·8)	11 (13·2)	21 (24·7)	9 (10·6)

Changes in arch status or footprint: boys (*Figure* 8.17)

Twenty-seven boys showed some change during the period of study. In six cases the arches became *more* apparent; all six of these children were under 6 years of age, and the alteration can be accounted for by the emergence of the longitudinal arch from the infantile flat foot. At 6 years old, or earlier, infantile valgus is no longer apparent, and the fat pad has disappeared.

In the remaining 21 cases, the arches became less apparent. Most of the boys were over the age of 12 when changes in pattern occurred; there appeared to be no correlation with fat percentiles. Fat percentile values were obtained by taking skinfold measurements in the following sites, i.e. over back of arm, over triceps.

It seems reasonable to suppose that, when footprint changes during the first six years, the arch may appear to become higher for physiological reasons; in early adolescence the arch may become less apparent,

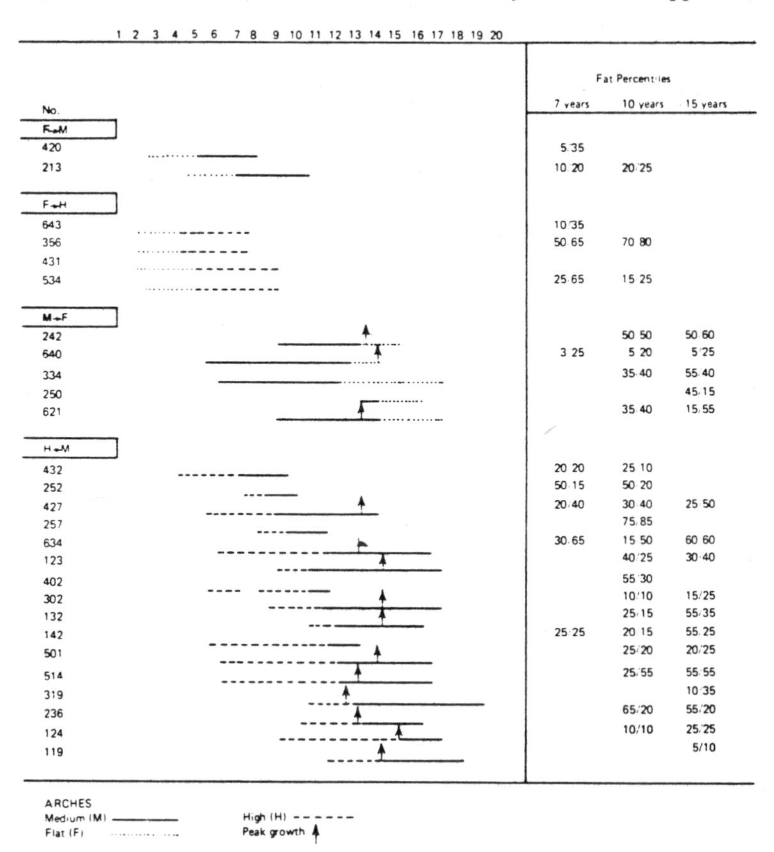

Figure 8.17. Changes in pattern of arch during study; boys

and this may be associated with the growth spurt; in middle school years, there is very little alteration in pattern.

Changes in arch status or footprint: girls (*Figure* 8.18)

Ten girls showed some change in footprint during the study. In 5 of them there was an apparent change in the height of the arch; like the boys, all these girls were under 6 years of age, and the change may be regarded as physiological in origin.

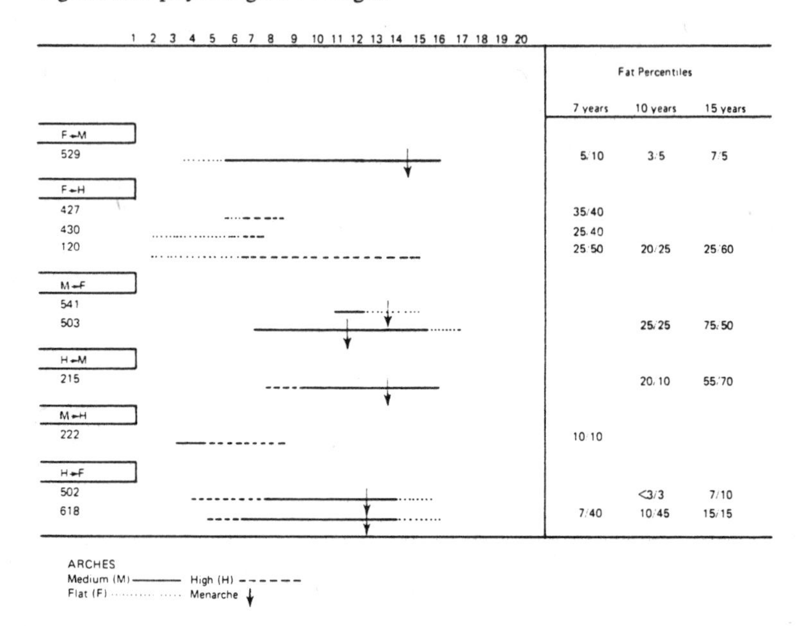

Figure 8.18. Changes in pattern of arch during study; girls

In the remaining cases, footprints appeared to be flatter than they were before. There is no definite pattern in these downward changes; in one case (503) fat percentile values had increased. There appears to be no correlation between the change in pattern and onset of menarche or peak of growth. Growth in width may obliterate the plateau space and give a flatter appearance.

Change in valgus heel pattern: Boys (*Figure* 8.19)

Twenty-one boys showed some change in valgus pattern; with the exception of two cases, all changes involved an increase in degree of valgus.

Of the nineteen who showed an increase in the degree of valgus, most of the boys were between the ages of 10 and 17; most boys seem to have developed valgus heel pattern during growth spurt. No well marked changes in pattern emerge. There is no definite sequence as in the longitudinal arches; this is not surprising, as valgus is a postural mechanism, used in time of need or stress. An example of increasing 'valgus heel' (ages 12–15) is seen in Figure 8.19.

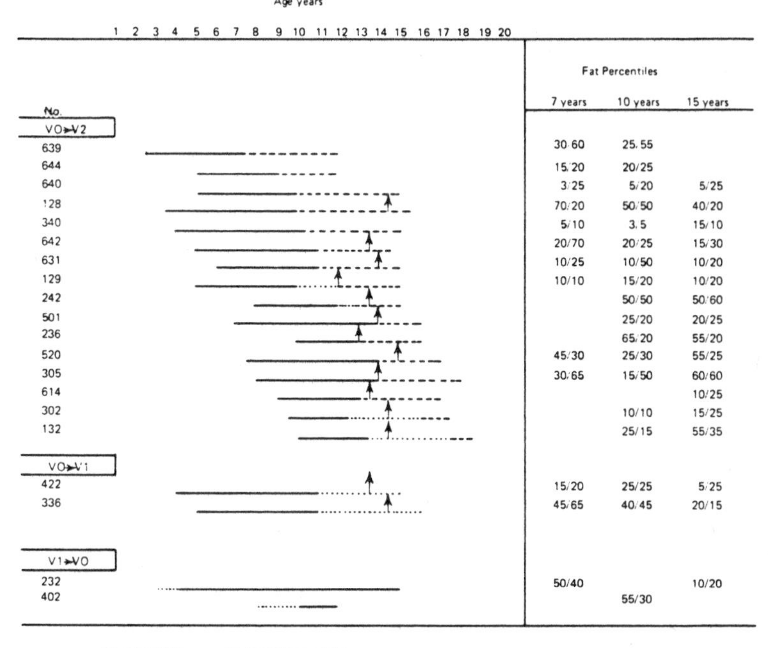

Figure 8.19. Changing degree of valgus heel during study; boys

Changes in valgus heel pattern: Girls (*Figure* 8.20)

Twenty-seven cases showed some change in valgus pattern; in twenty cases, valgus increased in degree. The age range differed from that of the boys (6½ to 14½). In general, changes began earlier, and this suggests a correlation with earlier onset of the adolescent spurt in girls. There appeared to be some positive correlation between fat percentile values and the occurrence of valgus heel.

The remaining six girls changed from 'mild valgus' to 'no valgus' during adolescent spurt.

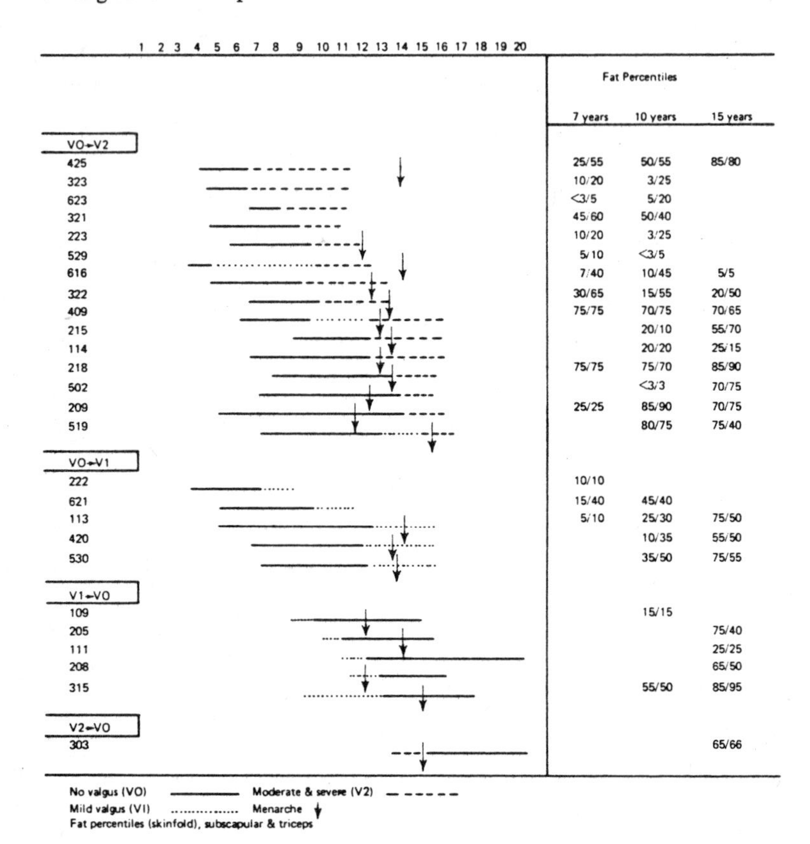

Figure 8.20. Changing of valgus heel during study; girls

CLINICAL NOTES

The infant's foot

The infant's foot appears to be very flat when he begins to walk, and the pedograph shows filling in of space XYZ; yet he has a well marked arch concealed by his fat and by his valgus posture. His postural reflexes are not fully developed, and he paddles around in almost total valgus for his first ambulant years.

According to Bruce and Walmsley (1938), who have investigated the fetal arch, the precartilaginous precursor of the foot is already arched in a manner very like that of the normal adult foot by the end of the third month of fetal life; changes which occur later both in the fetus and the infant are imposed on the original arched form. As postural reflexes develop, and fat decreased, walking assumes a more mature form, and the arch becomes apparent.

The foot with moderately high arch and valgus heel (Grade V2)

This is commoner in boys than in girls. The high arch in a long foot is difficult to manage, and the problem of balance is often solved by eversion of the heel. (Rated 2 valgus). This gives a characteristic foot, with internal malleolus descending slightly, and the lower end of the tibia, because it is nearer to the surface, showing as a sharp point through the skin. This valgus response appears to be quite a good solution to the problem, and should not be regarded as a defect. The bronze charioteer (fifth century B.C.) in the Delphi museum has feet of this type, with broad forefoot, high instep and short internal malleoli, with slight eversion of calcaneus and slight decent of the lower end of the tibia. We can only assume that this type of foot was not unusual amongst Olympian athletes, and perhaps we should remember this and treat valgus heels with more respect.

Children with mild valgus heels

The whole of childhood is a period of uneasy balance and experimentation, both with foot postures and with hip and shoulder postures; hence the wriggling and shuffling. There is no stable posture of rest to which the child can turn; he has no desire to rest! Small children standing barefoot will rarely plant a foot down and leave it there; they immediately begin to invert and evert their feet, to find a more satisfying posture, and may stand on the outsides of their feet curling up their toes. They are indeed, carrying out movements which we may later try to teach them as remedial exercises. It is rare to find a child under seven who stands still with no eversion, and this accounts for the large number of valgus heels in the Tables. These mild degrees of valgus are not usually reflected in the pedograph.

Children with more pronounced valgus

A high degree of valgus may be necessary if a very high arched foot is to be balanced; this can produce a very ugly foot with marked overhang of internal malleolus, tibia decending, navicular tubercle visible and sometimes separation at the talonavicular joint (*see Figure* 8.11).

The valgus feet may also be abducted. Such feet are likely to improve at puberty, when, owing to growth and changing proportions, the arch may be less high, and less valgus is necessary.

The adolescent foot

The foot shares in the general growth spurt. Growth in width is apparent, and more soft tissue is seen in the pedograph. A bridge may be seen in a previously high arch, converting it to a medium arch. The presence of more muscle and soft tissue is apparent everywhere. Valgus may increase a little, but it is often only a temporary measure, and rarely persists into the late teens.

Flat foot

In this study, we have considered only postural flatfoot. A congenital type exists in which the talus and calcaneus are connected by bone, but this is rare. There are also other causes of flat foot, which will be discussed here only briefly.

A foot which looks very flat is usually a valgus foot. In some cases eversion is so severe that medial border of calcaneus rolls over so that more of the inner border touches the ground; this can be demonstrated on the pedograph. In severe cases the space XYZ will be filled with faint inkprints, showing that soft tissue touches the ground intermittently but does not bear weight.

The best way of convincing ourselves and the parents that flat feet are postural and not due to 'dropped arches', is to make the child run and to inspect the foot from the medial aspect; in most cases the shape of the arches becomes apparent while the foot is off the ground.

CONCLUSIONS

(1) The arches of the foot rarely alter in pattern; in each individual, the shape is genetically determined.

CONCLUSIONS

(2) Valgus heel (sometimes erroneously called valgus ankle) is a postural device, which enables the child to redistribute weight when changes in body proportion occur. Unless it is severe, it cannot be regarded as a defect.

SUMMARY

In the early years of life, a medium or high arch emerges from the infantile 'flat' foot; the mid-growth spurt (6–7 years) probably affects the foot, and the child is able to take full advantage of the valgus-heel balancing mechanism during school years. When the foot grows during adolescence, valgus heel may be more marked for a time during the period of adjustment when extra weight-bearing is needed. Finally, by the age of twenty, in most cases valgus is no longer necessary and is rarely seen.

REFERENCES

Bruce, J. M., and Walmsley, R. (1938). 'Some observations on the arches of the foot and flatfoot'. *Lancet,* **2.** 656

Joseph, J. (1960). *Man's Posture: electro-mygraphic studies.* Oxford: Blackwell

Additional reading:

Roberts, W. H. (1962). 'Femoral torsion in normal human development'. *J. Bone Jt. Surg.,* **143,** 369

CHAPTER 9

9 Hallux Valgus

Hallux valgus means an outward or lateral deviation of the big toe.

It develops in childhood and, in Great Britain increases in incidence in both sexes (but more rapidly in girls) after the age of seven.

It may be progressive and may lead to bunions.

ANATOMY

First metatarsophalangeal joint and big toe (*Figure* 9.1)

The rounded head of the first metatarsal bone fits into the shallow cavity in the base of the proximal phalanx; flexion, extension, abduction and adduction of the big toe are said to be possible at this joint. The extensors of the big toe are inserted into the dorsal aspect of the bases of the two phalanges; the flexors are inserted into the plantar aspect of the base of the terminal phalanx (flexor hallucis longus), and with adductor and abductor hallucis into the plantar aspect of the base of the proximal phalanx.

The abductor hallucis is probably more concerned with steadying toes and with binding muscles into strong bases of support than with abduction of the big toe; it is quite well developed, and forms a fleshy elevation which extends forwards across the hollow of the instep; little is known of its real function.

The phalanges

The big toe has two phalanges; the remainder of the toes have three each. The *proximal phalanx* is broad and strong, but shorter than the others; the big toe, unlike the thumb, is not opposable. Hallux valgus is said to be present if the proximal phalanx deviates laterally from the direction taken by the metatarsal bone.

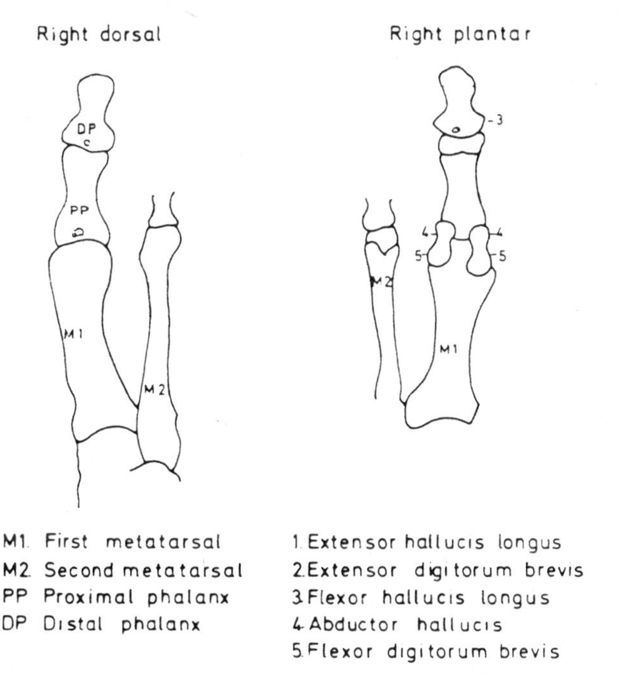

Figure 9.1. Metatarsophalangeal joint and big toe

The *terminal phalanx* is short and squat, and rather irregular in outline; the long flexor is inserted into the plantar surface, and the long extensor into the dorsal surface. Deflection of the terminal phalanx is difficult to measure owing to its squat irregular shape.

The metatarsal bones

The first metatarsal bone is shorter and thicker than the other four, which are long and slender. The bases of the metatarsal bones are heavy and their heads (with the exception of the head of the first metatarsal) are small and compressed laterally; their function is to assist in preserving the balance of the body.

EVOLUTIONARY ASPECTS

The human foot, as we know it, is the result of the specialization which occurred when man's ancestors left the trees for the ground (*Figure* 9.2)

It will be seen that, whereas the ratio of metatarsus to foot length is the same in chimpanzee, gorilla, infant and man, the tarsus occupies more than half of the foot length in man, and only one third of its length in apes. The phalanges in man are short, less than one-fifth of the length of the whole foot; those of the apes, on the other hand, occupy more than one-third.

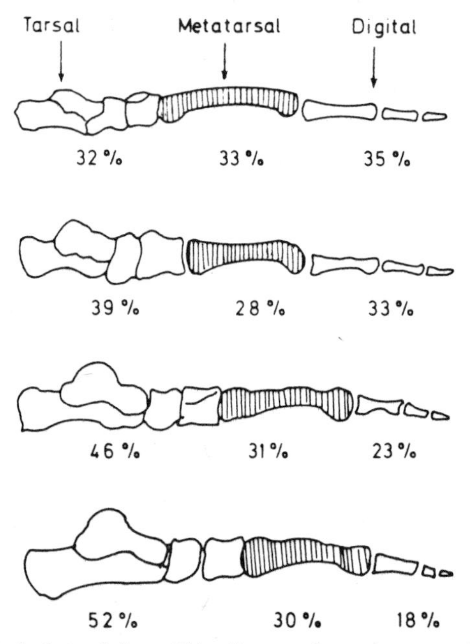

Figure 9.2. Evolution of foot. This diagram shows increase of the tarsal and decrease of the digital elements during the evolution of the higher primates. The change in proportions from infancy to adulthood is also shown

(Lake, 1936. Reproduced by permission of the publishers.)

Thus when man's ancestors left the trees for the ground, selection pressures favoured shorter phalanges as there was no longer any need for a grasping foot. The big toe was brought into line with the other toes. In order to cope with weight-bearing, the calcaneum increased in size and projected posteriorly, and the tarsus became more massive; the device of the arch enabled the foot to be planted firmly on the ground.

Individuals with mobile big toes may be more prone to hallux valgus than others whose toes are not so mobile, as such toes are less likely to produce a forefoot with a straight inner border or to follow the direction of the second metatarsal. In fact, infants with wide gaps between their toes and large metatarsal angles may be candidates for hallux valgus though this may not develop until the growth spurt is under way.

CAUSATION OF HALLUX VALGUS

Stamm (1957) states categorically that hallux valgus is always due primarily to a congenital defect in which the first metatarsal bone is inclined inwards, in the same way as the thumb is turned inwards in the hand. The comparison between the metatarsal bones in *Figure* 9.4 illustrates this point; the nine weeks fetus shows a wide angle between first and second metatarsal (*see* also *Figure* 9.3).

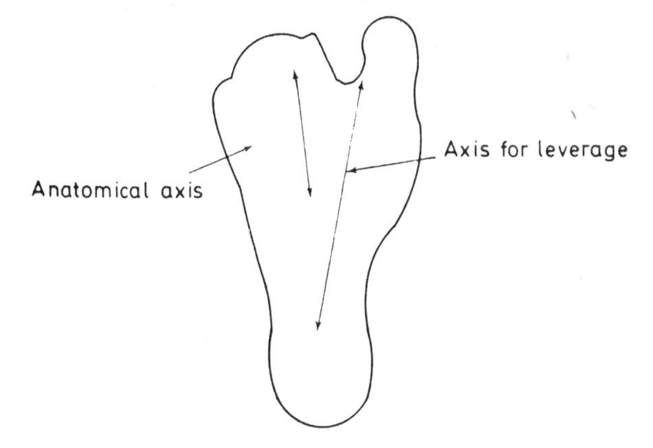

Figure 9.3. Foot of ten-months-old infant

Figure 9.5 (Stamm, 1957) shows (1) both toes and metatarsal in line, but inclined inwards; this is congenital. In (2) the big toe is pressed into line with other toes, and makes an angle with the metatarsal; in (3) the angle increases, and the deformity of the toe increases.

Many orthopaedic surgeons share Stamm's views, for example, Lake, (1952), Cholmeley, (1958), Hawkins, (1945), Ellis, (1951), Mitchell, *et al.* (1958), Rocyn Jones, (1948), Bonney and Macnab, (1952). Most treatment is based on the assumption that metatarsus primus varus is the primary factor, and many operative procedures include osteotomy of the first metatarsal, though various modifications have been made. Treatment must be carried out before growth of long

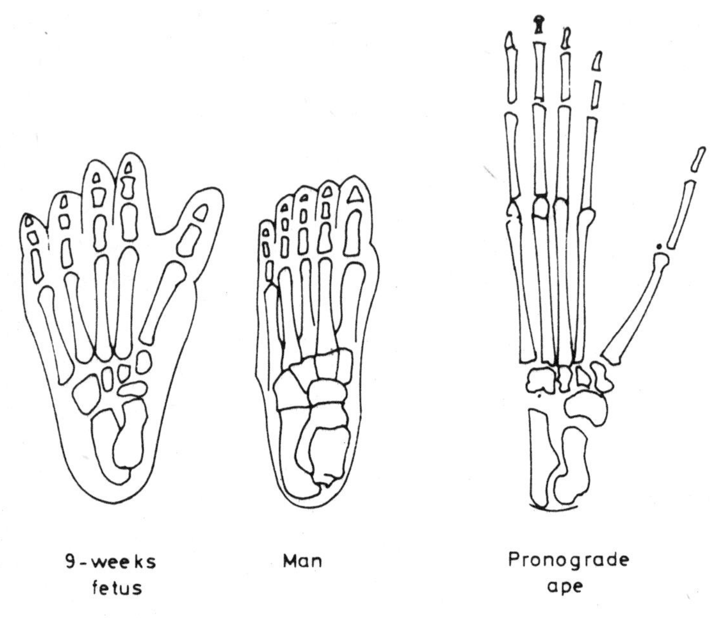

Figure 9.4. Metatarsal angles in fetus, man and pronograde ape (Lake, 1936. Reproduced by permission of the publishers)

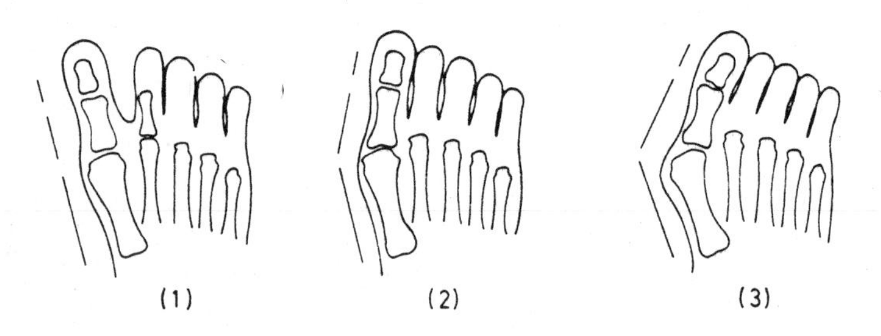

Figure 9.5. Stages in development of hallux valgus (Stamm, 1957. Reproduced by permission of the publishers)

bones has finished, and there must be no encroachment of articular surfaces. Bonney and Macnab (1952) made a critical survey of results of operative treatment of hallux valgus and hallux rigidus; most patients had large metatarsal angles, and they assumed this to be primary; they note however that, in 34 out of 54 feet, metatarsus varus recurred and, in some cases, the angle was larger than it had been before the operation.

METATARSUS PRIMUS VARUS

Metatarsus varus, or increase in the size of the angle between the axes of the first two metatarsals, may be suspected clinically in children when there is an obvious gap between the first and second toes; there is often some interdigital webbing, and the increased distance between the toes can be detected on the pedograph print.

Various figures have been given for the normal angle between the first two metatarsals. Ellis (1951) suggests 5.7 degrees for a control normal group; Hardy and Clapham give 8·5 degrees as metatarsal angle in control series. Mitchell *et al* (1958) suggest that angles of 10 degrees or more may be suspected of being abnormal.

Jones (1949) stresses the importance of the *total* divergence of the metatarsus, and describes an 18 mm embryo (7th week) in which the divergence of the first two metatarsals is 13 degrees and the total divergence is 40 degrees. He says that human divergence at all stages of development is a general splaying from the metatarsus, spread from the centre fan-wise. In the apes, by comparison, there is an isolated divergence of the great toe (*Figure* 9.4).

Lake (1952) suggested that hallux valgus might be instrumental in producing splay foot. As the forefoot is less sturdy than the rest of the foot the strain of the deformity might separate the first and second metatarsals, thus, the sequence could be hallux valgus–splay foot–increase in metatarsal angle.

Hardy and Clapham (1952), after measuring the intermetatarsal angle and the angle between the axis of the proximal phalanx and the axis of the first metatarsal in 1,850 school children, came to the conclusion that 'a congenital widening of the intermetatarsal may not be as important in hallux valgus formation as has been thought'. They noted that there was high correlation between the hallux valgus and wide metatarsal angle in adult controls, they suggest that these must be causally related but that displacement of big toe is primary, and causes widening of the intermetatarsal angle.

Hardy and Clapham in this cross-sectional study found a progressive increase in the displacement of the big toe with increasing age in both boys and girls. They found no significant increase in size of metatarsal angle in boys; there was however a progressive increase in girls. These results are interesting, though too much reliance cannot be placed on them as no allowance is made for differences between the sexes in developmental age; girls are on an average two years ahead of boys in skeletal age.

METHOD OF STUDY (87 BOYS AND 86 GIRLS)

Children's feet were examined six-monthly. Special note was taken of the posture of the inner border of the foot, the presence or absence of enlargement of the hallux (graded as mild, moderate and severe) and presence of a gap between first and second toes (rated as metatarsus varus). No measurements of the intermetatarsal angle were taken; all assessments of metatarsus varus were clinical. Pedographs were taken of all children six-monthly, as part of the general foot examination.

Rating of the foot for hallux valgus started some time after the main study had begun, and rating for metatarsus varus later still; for this reason available data were scanty, and only general trends will be discussed.

Cross-sectional studies

In *Figure* 9.6 the percentage of children with some degree of hallux valgus is plotted against chronological age. It will be seen that the percentage with hallux valgus increases in boys after the age of thirteen is reached; in girls the percentage increases earlier (owing to earlier growth spurt) and reaches a much higher level than that of the boys. The percentage is twice as high in girls as in boys by the time the girls are fourteen.

In *Figure* 9.7 the percentage of children showing metatarsus varus is plotted against chronological age. No clear picture emerges; there appears to be no definite connection with the growth spurt. The incidence is much higher in boys than in girls.

Long-term studies; hallux valgus and metatarsus varus

In these studies an attempt was made to find the relation, if any, of metatarsus varus to hallux valgus, and also to investigate the age of onset.

The following figures were obtained.

	Boys	Girls
Neither metatarsus varus or hallux valgus at any time	21	29
Metatarsus varus only	52	31
Hallux valgus only	1	4
Metatarsus varus *and* hallux valgus	13	22

Hallux valgus: *Age of onset* (years)		
7	1	2
8	0	2
9	3	3
10	3	4
11	1	6
12	3	4
13	0	4
14	2	1
15	1	0
16		

Long-term studies: comments

These figures are not very informative. Taken in conjunction with *Figures* 9.6, 9.7 it appears that:

(1) In boys, hallux valgus may occur at any time; in girls, it is more often related to the growth spurt.

(2) Hallux valgus in boys is more often related to splay foot, or to metatarsus varus, than it is in girls.

(3) In comparing these ages of onset of hallux valgus, allowance should be made for skeletal age.

FOOTWEAR: FOOT DEFORMITIES IN THE UNSHOD

Much has been written about the part played by footwear in the production of hallux valgus. During the growth spurt there is an increase in velocity of foot growth, and probably the long bones of the foot (metatarsals) increase in length at a greater rate than the tarsal bones. In many children there is a stage in foot growth when

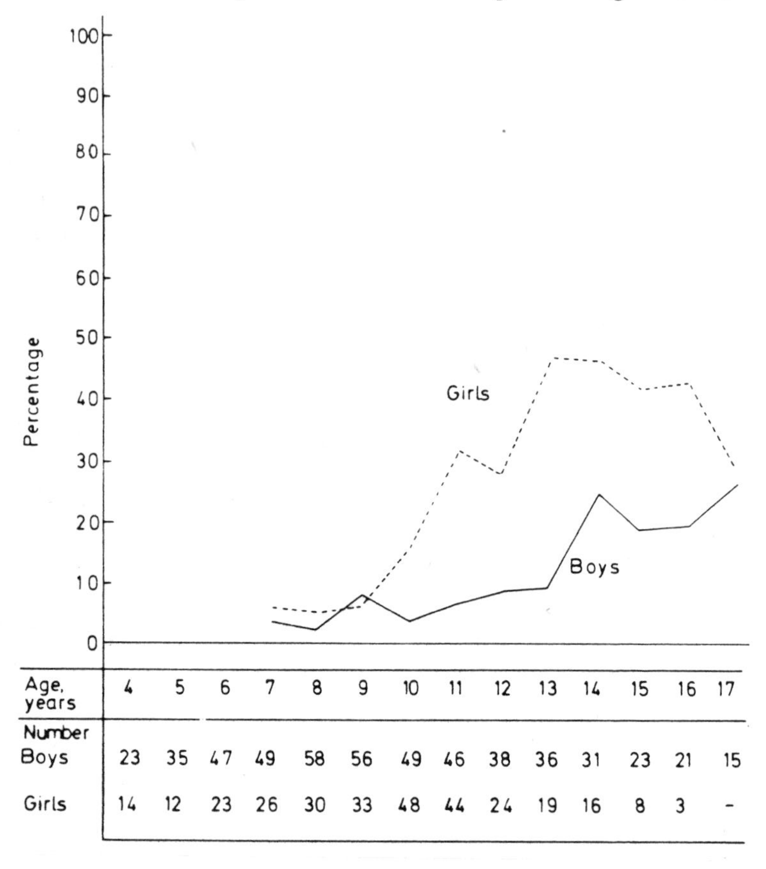

Age, years	4	5	6	7	8	9	10	11	12	13	14	15	16	17
Number Boys	23	35	47	49	58	56	49	46	38	36	31	23	21	15
Girls	14	12	23	26	30	33	48	44	24	19	16	8	3	-

Figure 9.6. Incidence of hallux valgus

the metatarsophalangeal joint is enlarged and the proximal phalanx is not deflected; at this stage, if shoes worn are not broad enough or long enough, toes may become crowded together and valgus may develop. Those who blame footwear for hallux valgus suggest that shoes with a straight inner border should be worn, so that big toes should have an opportunity to grow straight, in the same direction as the first metatarsal. The worse kind of shoe offered to our teenagers was the shallow low-heeled 'casual', which will not stay on the foot unless it is so short that the toes are crowded down towards the heel, and the brunt of pressure is borne by the big toe. Shoes with pointed toes are not quite so pernicious as these 'casuals', so long as the pointed toecap is beyond the end of the toes. The trendy squared medium-heeled shoes worn by teenagers at the present time are an improvement on the former shoe fashions.

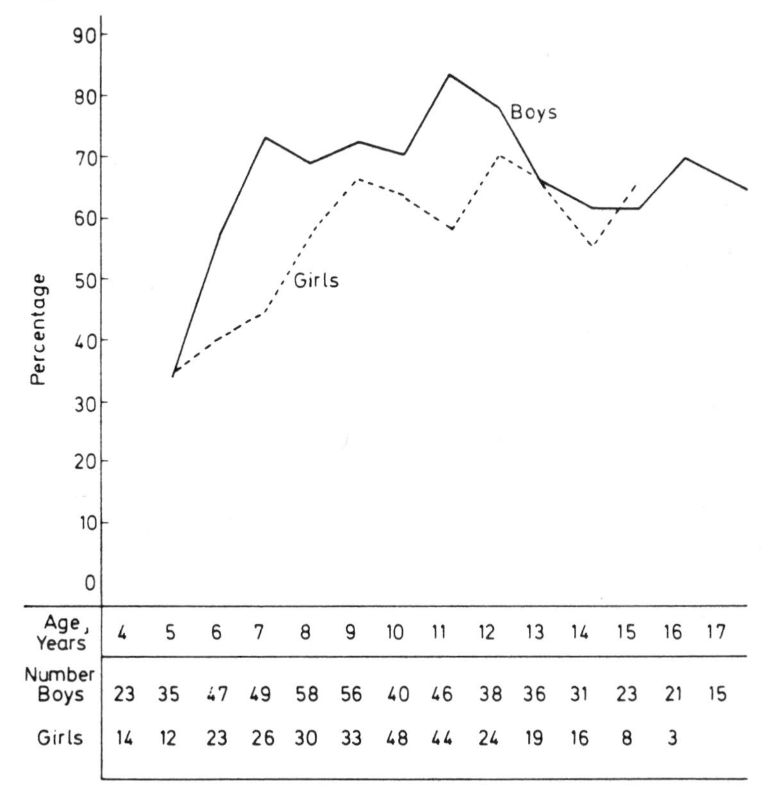

Age, Years	4	5	6	7	8	9	10	11	12	13	14	15	16	17
Number Boys	23	35	47	49	58	56	40	46	38	36	31	23	21	15
Girls	14	12	23	26	30	33	48	44	24	19	16	8	3	

Figure 9.7. Incidence of metatarsus varus

Though badly designed shoes may play a part in the production of foot defects, it cannot be assumed that there would be no deformities if shoes were always well designed, or were not worn; common foot deformities may occur in those who go barefoot.

Barnicot and Hardy (1955) noted that an occasional deviation measured by hallux angle was found in West Africans; they studied footprints taken with Scholl's pedograph (*Figure* 9.8). They noted also that, in general, hallux angulation increased with age. Wells (1931) noted a lateral displacement in a South African Bantu, and other isolated examples have been described.

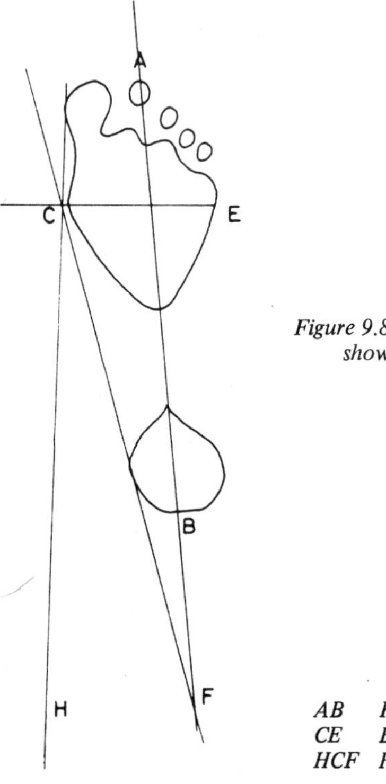

Figure 9.8. Pedographic tracing showing hallux angle

AB	*Foot length*
CE	*Breadth*
HCF	*Hallux angle*

Shine (1965) studied the natives of St. Helena. The majority were unshod. Some of them, both men and women, wore round-toed, low-heeled, laced-up shoes. The mean hallux angle was calculated, and a figure greater then 15 degrees was noted in 16 per cent of men, and in 48 per cent of women wearing shoes, but in only 2 per cent of barefoot natives of both sexes. Thus shoe-wearing showed a highly significant association with hallux valgus, and the difference in incidence in the sexes is apparent.

Shine notes that hallux valgus was absent in some families who wore shoes, and present in others who went barefoot; these cases may have a genetic basis.

CONCLUSIONS

What is the cause of hallux valgus? Hardy and Clapham (1952) conclude that the displacement of the toe is primary and causes widening of the metatarsal angle. In view of the high correlation coefficient in adult controls (0·7) between hallux valgus and intermetatarsal angle, they found it difficult not to conclude that they were causally related.

An important part of Hardy and Clapham's study deals with the distribution of the size of the intermetatarsal angle in age groups. Though they take no account of developmental age, it is obvious that the girls show a marked increase in the intermetatarsal angle between the ages of 10 and 11. Boys are beginning to show an increase two years later (as we should expect). If these measurements had been continued till growth had nearly ceased the boys, too, might show an increase in intermetatarsal angle with increasing age.

What then is the association of the intermetatarsal angle with hallux valgus? A large number of surgeons, notably Stamm (1957), suggest that there is an evolutionary explanation: the big toe has lost its mobility in favour of stability. Before deflection is apparent, the joint shows marked increase in size; at this stage deviation can begin, and footwear should be most carefully selected.

Can we blame unsuitable footwear for hallux valgus? Footwear is not the only cause; badly designed shoes, and certain skeletal patterns together, form good material for the production of hallux valgus and for bunions in later life. Yet we must not forget the unshod, who from time to time manage to produce hallux valgus, but never bunions, as far as is known.

The genetic factor cannot be disregarded. In the growth study hallux valgus was seen to develop in several members of the same family, including a pair of identical twins with identical halluces.

To complete the picture, some families of islanders who have never worn shoes have developed hallux valgus and some who have always worn shoes have never had hallux valgus.

SUMMARY

It is probable that there is a congenital skeletal pattern that favours the development of hallux valgus, and that hallux valgus may be prevented from developing if suitable footwear is worn. Children without this skeletal pattern are unlikely to develop hallux valgus, even if shoes are not entirely suitable.

The orthopaedic surgeons who support the congenital metatarsus varus theory advocate surgical treatment in the form of metatarsal osteotomy. Cholmeley (1958), however, says 'Metatarsal osteotomy may be considered meddlesome surgery in children and adolescents'. Splints and exercises are of doubtful value; shoes with a straight inner border might prevent deformity, but most children cannot be persuaded to wear them.

REFERENCES

Barnicot, N. A. and Hardy, R. H. (1955). 'Position of hallux in West Africans'. *J. Anat.,* **89**, 355

Bonney, J. L. W., and Macnab, I. (1952). 'Hallux valgus and hallux rigidus; critical survey of operative results.' *J. Bone Jt Surg.,* **34B**, 366

Cholmeley, J. A. (1958). 'Hallux valgus in adolescents.' *Proc. R. Soc. Med.,* **51**, 90

Ellis, V. H. (1951). 'A method of correcting metatarsus primus varus.' *J. Bone Jt Surg.,* **33B**, 415

Hardy, R. H., and Clapham, J. C. R. (1952). 'Hallux valgus: predisposing causes'. *Lancet,* **1**, 1180

Hawkins, F. B., (1945). 'Correcting hallux valgus by metatarsal osteotomy.' *J. Bone Jt Surg.,* **27**, 387

Jones, F. W. (1949). *Structure and Function of the Foot.* 2nd ed. London: Bailiere, Tindal & Cox.

Lake, N. C. (1936, 1952). *The Foot.* London: Bailiere, Tindal & Cox.

Mitchell, C. L., Fleming, J. L., Allen, R., Glenney, C., Sanford, G. A. (1958) 'Osteomy and bunionectomy for hallux valgus.' ***J. Bone Jt Surg.,*** **40a,** 41

Rocyn Jones, A. (1948). 'Hallux valgus in the adolescent.' *Proc. R. Soc. Med.,* **41**, 392

Shine, I. B. (1965). 'Incidence of hallux valgus in a partially shoe-wearing community.' *Br. med. J.* **1**, 1648

Stamm, T. T. (1957). 'The surgical treatment of hallux valgus'. *Guy's Hosp. Rep.* **106**, 273

Wells, L. H. (1931) *Am. J. phys. Anthrop.* **15,** 185

Additional reading:

Emslie, M. (1939). 'Foot deformities in children'. *Lancet,* **2**, 1261

James, C. S. (1939). 'Footprints and feet of natives of Solomon Islands.' *Lancet,* **2**, 1390

Appendix

DIFFERENTIAL DIAGNOSIS

A list is given below of the disorders mentioned in the text. In a book of this kind there is little place for detailed differential diagnosis; a list of conditions, many of them rare ones, may however be of interest.

The interested reader is referred to *Orthopaedics in Infancy and Childhood* by G.C. Lloyd-Roberts (1971), London: Butterworth.

The following approach to preliminary sorting may be of practical value:

A mother may want advice about her child's knees. No abnormality is found on examination, but the mother is worried. She should be reassured that no treatment is necessary, but there should be no reassurance without explanation (*The Child and his Symptoms*, by J. Apley & R. MacKeith, 2nd Ed. (1968) Oxford: Blackwell) and explanations may need to be repeated. The mother may be asked to bring the child to the out-patient clinic for a few months. This physiological type of knock knee will get better whatever we do. The child may (though it is unlikely) have an abnormal varient e.g. Blount's disease (*see* text) and may need surgical or medical treatment (*see* under Rickets) if a disease process is known to exist.

DIAGNOSIS

LIST OF DISORDERS

Name of disorder (normal variant)	Differential diagnosis from:
1. Mobile kyphosis	Structural kyphosis Poliomyelitis Muscular dystrophies Scheuermann's disease
2. Lordosis	Congenital dislocation of hip (bilateral)
3. Round shoulder	Congenital undescended scapula Winged scapula Klippel-Feil syndrome Scapulae fail to descend
4. Mobile scoliosis	Structural scoliosis with skin pigmentation in Von Recklinghausen's disease.
5. Bow legs	Persistent bow legs Achondroplasia Rickets (nutritional or metabolic) Blount's disease Osteogenesis imperfecta Spondylo-epiphyseal dysplasia
6. Knock knee, Genu varum	Associated with obesity (gross) Secondary to: Rheumatoid arthritis Renal or metabolic rickets Myelomeningomyocele
7. Flat foot	Spina bifida occulta Benign hypotonia Cerebral palsy Peroneal or spasmodic flat foot Congenital tarsal abnormalities

Index